The Hamlyn Guide to
Plant Propagation

The Hamlyn Guide to
Plant Propagation

Romneya hybrida

*edited by Susanne Mitchell
and Barbara Haynes*

Hamlyn
London · New York · Sydney · Toronto

Acknowledgements

Most of the black and white pictures first appeared
in *Pictorial Gardening.*

Additional black and white pictures
by Robert Corbin and John Cowley.

Colour photographs by Pat Brindley,
The Hamlyn Publishing Group, Robert Pearson,
The Harry Smith Photographic Collection and Michael Warren.

Jacket illustrations by Fiona Butler
Line drawings by
The Hayward Art Group and Norman Barber.

Some of the material in this book has previously
appeared in *Pictorial Gardening*, first published
by The Hamlyn Publishing Group Limited
in a revised edition in 1969

First published in this form in 1982 by
The Hamlyn Publishing Group Limited
London · New York · Sydney · Toronto
Astronaut House, Feltham, Middlesex, England

Third impression 1984

Phototypeset in England by Photocomp Ltd
in Monophoto Apollo 645 in 9 on $9\frac{1}{2}$pt and 11 on 12pt
Printed in Yugoslavia

ISBN 0 600 30516 3

Printed in Yugoslavia

Contents

Introduction

One of the most fulfilling aspects of gardening comes from growing plants you have raised yourself, either from seed, cuttings or by some other method of propagation. Most of the techniques required are simple, involve little in the way of equipment and offer a high percentage of success – provided the instructions given are followed.

For the person who wants to propagate quite a lot of plants, however, a propagating case will prove to be an invaluable item of equipment. There are many different types available but most work on the principle of having a thermostatically controlled, electrically heated base tray. This can be either covered with soil and used for direct sowing or cutting insertion or, depending on size, may hold a number of seedtrays. A clear plastic cover fits over the tray and retains the warmth around the plants. The great advantage of a propagating case lies in the fact that the slightly higher temperatures needed for faster rooting and germination can be confined to a small area and will result in savings on fuel costs. Other unheated

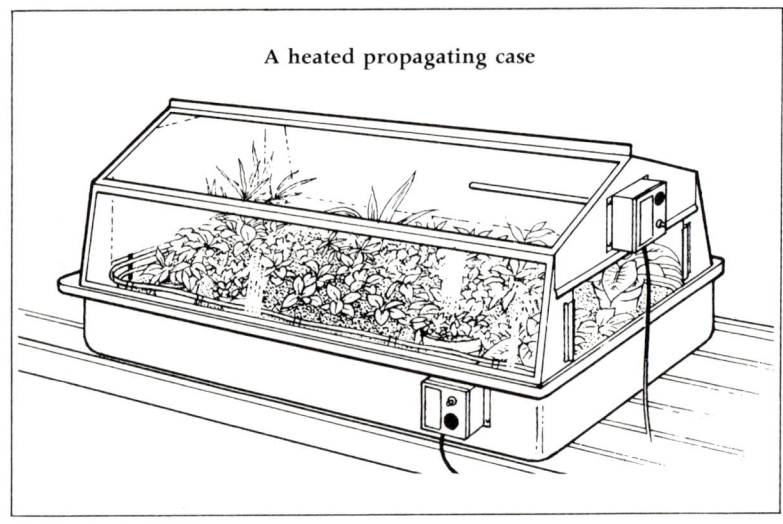

A heated propagating case

propagating cases are available, which rely on the increased warmth produced in a small, closed area for their effectiveness, although there is one type which slips over a house radiator and uses this as a source of heat. Alternatively, if you have only a few cuttings, the simple trick of placing the pot or tray in a large plastic bag will often suffice.

TOOLS

A sharp knife is a first requirement, secateurs, clean pots and trays, a sieve, labels and a dibber (small piece of wood with a pointed end – somewhat similar to a pencil) are all more or less indispensable. A pair of tweezers may be helpful, and an aid to pricking out made from a wooden plant label notched at the tip to form a small two-pronged fork.

COMPOSTS

One of the most essential requirements for successful propagation, together with scrupulously clean pots and trays, is the compost or soil mix used for sowing the seeds or insertion of the cuttings. When working indoors never use garden soil, which often harbours diseases and pests, instead always buy a specially formulated mix for the job. This may be either one of the John Innes seed or potting composts or one of the soilless equivalents. Not only will the texture of these mixes encourage fast rooting and successful germination but the correct amount of fertilizer has already been included to get the plants off to a good start.

It is sometimes necessary to lighten the compost and make it more free draining by the addition of sand, and this should always be sharp or silver sand and never the kind used by builders. Propagation carried out in the garden generally uses garden soil although this may be supplemented by sand or a compost mix.

TYPES OF REPRODUCTION

Plants reproduce themselves in two ways, either sexually or asexually. Sexual reproduction requires the production of sex cells, pollen and ovules, and the subsequent fertilization of the ovules by pollen from either the same flower or a different flower of the same or similar species or variety.

Like all cells, the sex cells carry genes which are responsible for the various characteristics, such as colour, number of petals, height and scent, exhibited by the plants and it is the recombination of the genes at fertilization which brings about variations in these characteristics in the offspring. This, very simply, is the basis of plant breeding. For example, the double-flowered varieties have been bred from species which, due to genetical changes, started to produce more than the normal number of petals. It is mostly the species which can be reproduced successfully from seed by the gardener, many named varieties (with the exception of some vegetables and annuals) cannot be reproduced from seed saved from the plant itself since this will often give rise to a wide variation in the young plants germinated from this seed. And much of the variation will result in plants which are inferior in many of their characteristics to the parent.

There is a curious phenomenon used in plant breeding which is known as hybrid vigour. This term describes the offspring from a cross between two selected varieties which turn out to be very uniform in appearance, are larger and have a better cropping ability than either of the parents. These offspring are known as F_1 and seed labelled as such is offered in nurseryman's catalogues. However, successive crosses made between members of an F_1 generation show a continual decline in vigour and uniformity so it is not worth saving the seed for growing another year. Each F_1 generation must be reproduced by the nurseryman from the same initial cross. Not all plants exhibit hybrid vigour.

It sometimes happens that a mutation or accidental change in a gene causes part of a plant, or indeed the whole plant, to develop different features such as flower colour, number of petals, height or habit. This gives rise to what is known as a sport and it can be a useful factor in plant breeding. Many sports if produced in a species will breed true from seed, in hybrids they may be perpetuated by vegetative reproduction. Variegation in leaves is often the result of mutation which brings two or more different kinds of tissue together and is one of the main reasons for striped or speckled leaves. This can be an unstable mutation because the leaves will sometimes revert to plain coloration again in less than ideal growing conditions.

For the most part, the gardener is more concerned about obtaining good results with seed bought from a nurseryman than by collecting and sowing his own. But in these days of ever-increasing costs it is worth trying one's own seed production for some of the annuals and vegetables as long as the variety is not an F_1 hybrid. Gather the seeds from good, healthy plants on a sunny day before the pods open. Spread them on paper in a warm place to finish ripening and then clean them by blowing the dust and old seed heads away. Pack in paper bags, label them and store them in a dry, cool place where they will be safe from the ravages of mice.

Asexual or vegetative reproduction describes any means of propagation which uses a part of the plant and, with few exceptions, reproduces the parent plant identically in all characteristics. The parts of the plant suitable for asexual reproduction are stems, leaves, buds, roots, bulbs, corms, rhizomes, offsets and tubers. These are the methods of propagation used for the majority of garden plants.

One very important fact to bear in mind is that whichever of these methods is used the parent plant from which the material is taken must be healthy and growing well. It is of no use whatsoever to take, for example, a weak, etiolated cutting from a plant growing in less than ideal conditions or affected by pests or diseases and expect it to root successfully and produce a sturdy plant of good habit. So always select the best parent plant available as a source of material.

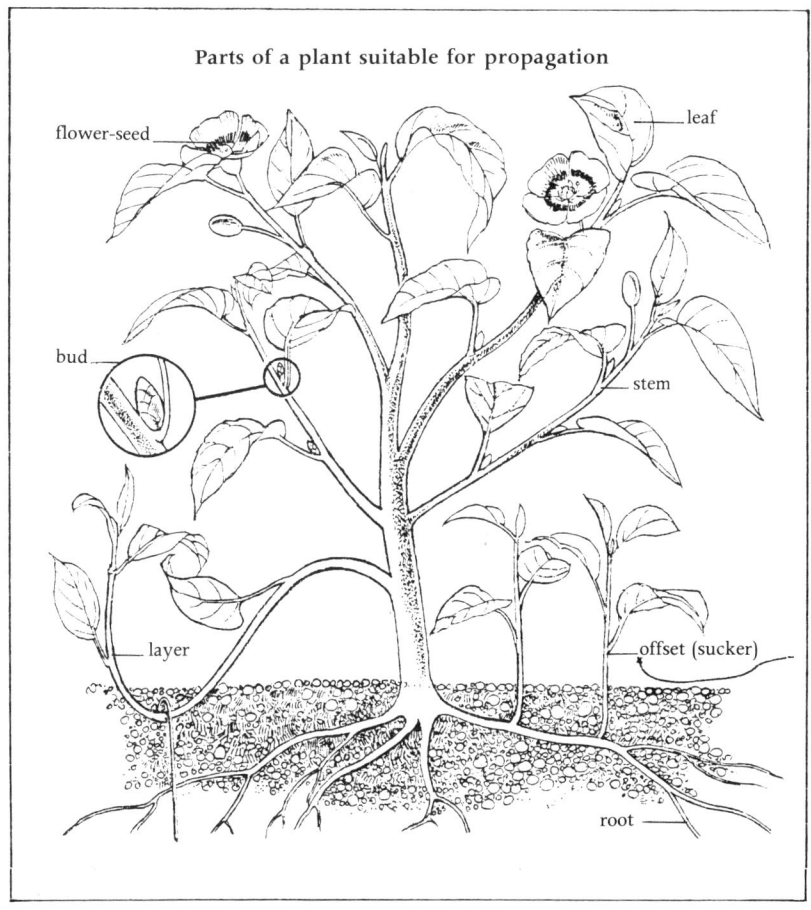

Parts of a plant suitable for propagation

flower-seed

leaf

bud

stem

layer

offset (sucker)

root

Sowing

Seed is probably the most common and perhaps the simplest method of propagating a plant. However, in order to meet with success a number of points are worth bearing in mind. A seed needs air, warmth and moisture in order to germinate, and the seedling, once it emerges, will need light as well if it is to make sturdy growth. Seeds from different plants vary greatly in size and appearance and it is these points that govern how they should be sown. A general rule to follow is to cover a seed with its own depth of soil. In practice this is often not as easy as it sounds.

Very fine seeds should scarcely be covered. These are scattered over the soil surface and then just lightly firmed in with a presser or piece of wood to make sure they are in contact with the compost. A useful aid is to mix fine seed such as begonia with a little sand to ensure that the distribution is even. Seed should always be sown thinly and evenly to facilitate pricking out and prevent over-crowding and its associated problems. Larger seeds such as lettuce or aquilegia can be covered with a light sifting of soil or compost. Seeds large enough to be handled individually like runner beans or sweet peas can be sown one to a pot rather than broadcast over the compost.

Seeds with hard coats, like sweet peas, will germinate quicker if this coat is nicked with a penknife well away from the hilum or eye. Others like cyclamen or the fleshy seeds – peas and beans – can benefit from a twenty-four hour soaking in water.

A seed showing the hilum or eye

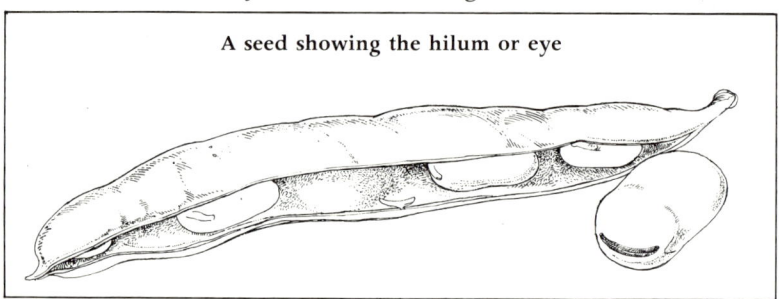

Seeds which are slow to germinate or are expensive to buy can be pre-germinated in a gel or on damp blotting paper kept in a jar in the airing cupboard or some other warm, dark place. They should be inspected daily and as soon as germination has taken place they must be removed and sown in compost or in the soil outside. Those sown in gel can be squeezed along the length of a drill using a polythene bag with one corner snipped off or a bottle manufactured specially for this purpose. The flow of gel can be controlled and ensures the seeds are correctly spaced so thinning out will be unnecessary. Pre-germinated seeds of difficult subjects can be bought from seed companies. They are expensive but save time and the cost of providing the often very exact conditions required for success. This can involve the need for very high temperatures initially.

A propagating case is a useful aid to germination. It can be fitted with a heated base to produce a controlled heat over a small area. On a larger scale a garden frame can have soil-warming cables installed for getting vegetables and half-hardy annuals off to an early start.

Many shrubs, such as cotoneaster, holly or pyracantha, produce berries or other seeds with a fleshy covering over a woody seed coat. These seeds need to be subjected to a process known as stratification before germination can take place. This involves the seeds undergoing a period of cold followed by a period of warmth and the whole process can take up to eighteen months or more. In contrast quick-germinating seeds like radish, mustard and cress take just a few days.

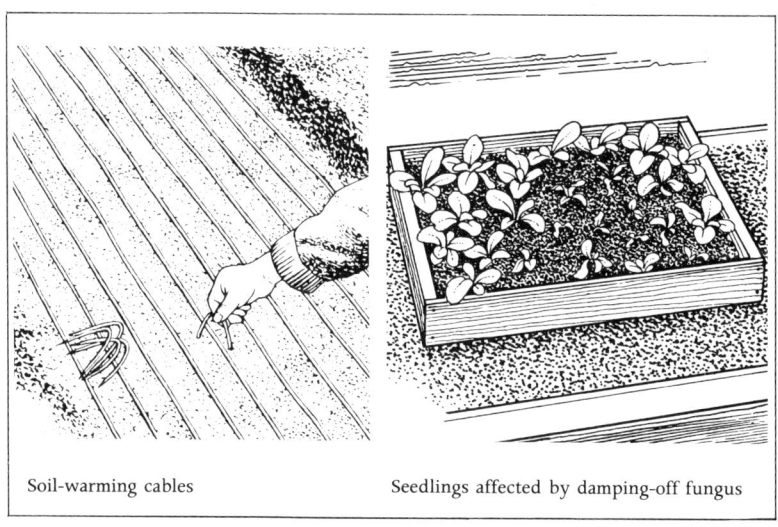

Soil-warming cables   Seedlings affected by damping-off fungus

It is advisable to use a special seed compost when sowing in containers. There are two main kinds available – one based on peat and the other on loam. Of the latter type the John Innes formula is the best known. These specially-formulated composts ensure the germinated seedling has just the right amount of fertilizer to satisfy its needs. This is why seedlings should be transplanted into a richer mixture as soon as they become large enough to handle if they are to continue to make strong healthy growth. It is important always to use fresh compost as old, stale compost may harbour pests or diseases and, of course, it will no longer be a balanced growing medium.

The most common disease of seedlings is damping off fungus. This rots the base of the stem causing the seedlings to keel over and die. It can be prevented by watering the compost with a solution of captan, zineb or Cheshunt compound made up according to the instructions on the packet. Stale air and overcrowding encourage the disease, so sow thinly and thin seedlings as soon as they can be handled, also ensure ventilation is adequate.

Sowing outdoors requires the preparation of a fine tilth and details are given on page 28. It is important that the ground should be consolidated as loose puffy soil can lead to problems later on, especially with onions and brassicas. There are a number of soil-borne pests and diseases which can attack germinating seeds and seedlings. In answer to this some seeds can be bought dressed with chemical deterrents. The latter usually come as dusts and can be applied following the manufacturer's recommendations. A calomel dressing protects onion seed against white rot and onion fly, and peas and beans can be dusted with lindane to prevent attack by wireworms, leatherjackets and other soil organisms.

You may be tempted to raise plants from seed you have collected or been given by friends, but do remember that many plants have undergone long and complex breeding programmes to induce a higher yield or an extra large flower. These are the F_1 hybrids which will not grow true from collected seed and are therefore bound to be a disappointment. Seed gathered from named varieties, too, can give rise to a wide range of offspring, often quite different from the parent plant. However, species can be raised from seed quite successfully and it can be fun, as well as rewarding, to experiment.

Seeds not used up in one season, or that have been collected from plants, can be kept until the next suitable sowing time in a cool, dry place but never in a polythene bag. There are some seeds, such as meconopsis, that lose their viability very quickly and should, therefore, be sown as soon as they are gathered. Generally speaking,

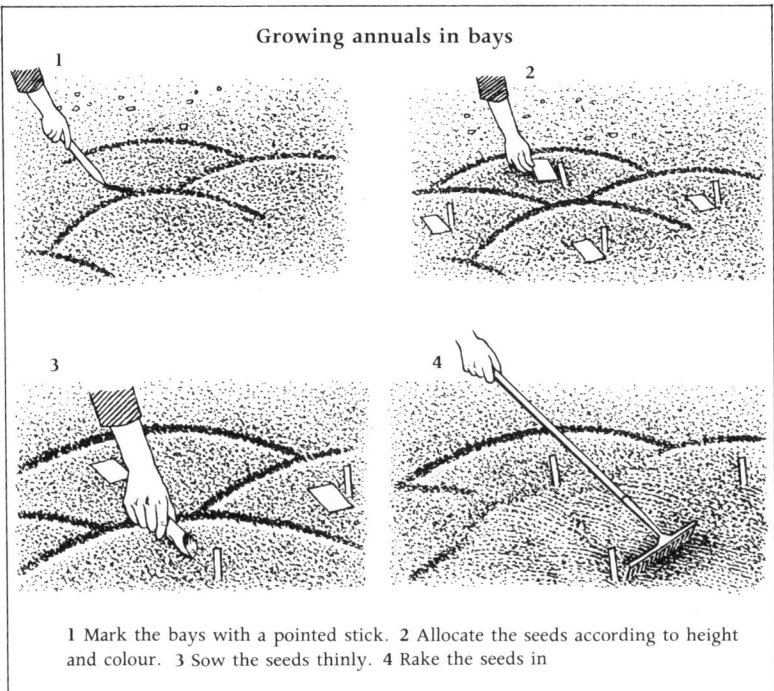

Growing annuals in bays

1 Mark the bays with a pointed stick. 2 Allocate the seeds according to height and colour. 3 Sow the seeds thinly. 4 Rake the seeds in

seeds can be kept for more than a year, and even after two or three years the percentage that germinate will still make sowing worthwhile. However, it is advisable to sow a little more thickly to prevent gappy rows due to the reduced viability of old seed.

Sowing information and times are always given on the seed packet but it is worthwhile bearing in mind the following provisos. In colder parts of the country be prepared to sow rather later if conditions at the time advised are not suitable. When sowing outdoors allow time for the earth to warm up and never sow when the ground is frosted or waterlogged or when soil adheres to the tines of the cultivator or rake. If sowing early indoors or in the greenhouse be certain the necessary high temperatures and good light can be supplied and maintained until planting-out time. Therefore, if greenhouse space is restricted don't be in too much of a hurry to get things started. Young plants standing in low light become weak and etiolated and will not do well when planted out.

To economise with sowing compost and help retain moisture, cover the base of the tray with ½-in (1-cm) layer of damp peat. Then fill tray with either a soil-based or peat-based seed compost.

Firm the compost down well using the fingers to ensure there are no pockets of air present.

Obtain a level surface by giving a final firm with a presser. This is a simple tool to make out of a flat piece of wood with a smaller piece stuck or screwed on to act as a handle.

Water the compost well using a watering can fitted with a fine rose. Allow any excess water to drain away.

Broadcast seed thinly over the surface of the compost.

Cover the seeds with a thin layer of sieved compost and lightly firm again with the presser. Remember seed should only be covered with their own depth of compost.

Finally label the seeds. The trays can be covered with a sheet of glass and a layer of newspaper to encourage germination. As soon as the seeds have germinated the glass and newspaper should be removed, so check the trays frequently for the first signs of green showing. Keep the developing seedlings in the lightest position possible.

For sowing small batches of seed use a 3½- or 5-in pot, a half-pot being the most suitable container. If the pot is made of clay, then cover the drainage holes with pieces of crock. Pots must be clean, and if not new, washed well beforehand.

Cover the crocks with about 1 in (2.5 cm) well-rotted compost or turf fibre. Alternatively peat may be used. Press this down fairly firmly, then fill the pot with J.I. seed compost or a peat-based equivalent.

After filling the pot with compost and firming evenly, level the surface using the base of a small pot or a presser. Leave ½ in (1 cm) between the top of the pot and the compost as watering space.

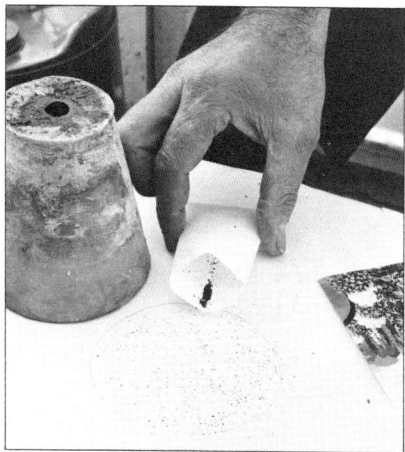

Small seeds, which are not easy to see, should first be tipped out on a sheet of white paper with the size of the pot to be used for sowing marked out on it. The amount needed to cover thinly the area of the pot is then poured into a small container for the actual sowing.

With small quantities of seed, two kinds can be sown in the same pot, with a label placed across the centre to act as a division. Label each kind of seed clearly with its name and date of sowing.

Cover the seed with finely sieved soil. Large seeds should be covered by about $\frac{1}{4}$ in (0.5 cm), smaller seeds more shallowly. Very small seeds should not be covered, but simply pressed into the surface.

After sowing, water the pots carefully with a fine-rosed watering can or stand them in a bowl of water so that the moisture soaks upwards through the drainage holes. Avoid over-saturation, particularly with small pots.

Sweet pea seeds or other large
seeds can be sown singly in
multi-pots. These plastic
containers are light and easy
to handle, and save space. The
young plants have space to
grow on in these and pricking
out is unnecessary.

Bottomless black paper tubes
can be used for many
vegetable and flower seeds.
When the tray load is
complete, pour compost over
the pots until they are full;
firm and then sow seed or
prick out seedlings. At a later
date, each plant can be
removed from the box without
root disturbance.

The Jiffy pot range is now
quite wide and as many as
sixty of these peat pots can be
fitted into a standard seed
tray. They are easy to handle
and require very little
compost to fill them.
Transplanting is simple as the
peat pots can be easily
separated. No root disturbance
takes place and there is no
check to growth.

Special blocking compost is available from garden centres and most horticultural sundriesmen. This must be thoroughly moistened following the instructions on the plastic sack. It is necessary to make it slightly damper than is normally used to ensure the blocks retain their shape.

The soil-block mould can also be bought from garden centres. Lift the plunger and press down firmly into the compost.

Press the mould filled with compost down firmly on a hard surface and push the plunger so the block is left standing free. Space the blocks evenly on a board or the base of a seedtray so they can be moved easily.

Sowing or pricking out of seedlings can then proceed. One seed or seedling is dropped into the depression in each block. Water carefully using a fine rose on the can and ensure that the compost never dries out.

Jiffy 7's are small discs of compressed peat encased in netting. Stand them in a shallow tray of water or a seedtray lined with polythene and partly fill with water.

After a short time the Jiffy 7's will swell to several times their original size.

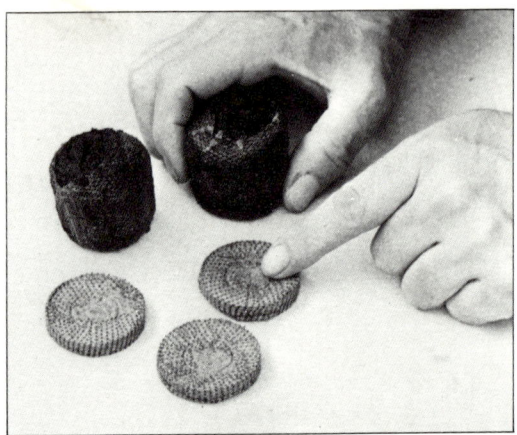

Place them in a seedtray or a purpose-made plastic tray and sow one seed to each using a dibber. After germination has taken place the seedling can easily be planted out in its peat block as the roots can grow out through the netting.

Where flower seeds are allowed to ripen on the plant gather the 'heads' when quite dry and store in an airy place. The seed heads can be put temporarily in a polythene or paper bag which also prevents seed being lost.

If the seed is to be sown immediately after gathering, shake the heads over a prepared container. With many flower seeds, it is sound practice to sow as soon as the seed becomes ripe.

Seeds may be shaken out onto a piece of white paper and then tipped into a small packet for ease of sowing. If they are not for immediate use, store the seeds in a paper bag or a tin in a dry, cool place until required and label the container.

Seeds may also be sown from the hand: take the seed between thumb and finger and space evenly over the surface of the pot.

Germination of hard-coated seeds can be improved and speeded up if they are exposed to frost or a period of cold temperatures before sowing. Many seeds of trees and shrubs benefit from this treatment, including rose heps. Start by putting a layer of sharp sand into the bottom of a pot or seedtray. Follow this with a layer of heps, then another layer of sand. Repeat these layers until the pot is full, always finishing with a layer of sand.

The seeds must be stratified out-of-doors in an exposed position. They can be protected from flooding by covering each pot or tray with polythene and from rodents, especially mice, with some chicken wire.

If a number of different types of seed need to be stratified it is a good idea to sow each variety in a separate, clearly labelled pot. Group the pots or trays together and cover them all with a length of chicken wire, securely pegged down all around the edge to prevent mice from stealing the seeds.

When small quantities of different cacti are being sown, divide the pot into three or four sections, with pieces of celluloid labels pushed into the compost. Number the sections for identification.

Tip the seed into a tablespoon and gently shake the seed from this, spreading it as evenly as possible. Cactus seed germinates best if left on the surface.

After watering the pots, which is best done by standing them in a container of water until moisture is seen on the surface, drain and place them in a propagating case and cover with a piece of glass and a sheet of black paper.

When pricking out seedlings, which resemble tiny beads, use an old plastic knitting needle with a cleft cut into it. Handle with care, as broken roots will cause the seedling to die.

A small batch of seed may be sown in a pot or seedtray in August in J.I. seed compost or a soilless compost, or in April under glass. Lift the seedlings when large enough to handle, taking care to keep the root system intact.

Prick out six seedlings to a seedtray in either J.I. potting compost No. 1 or a soilless compost. Delphiniums make a large fibrous root system, and by pricking out only six to a tray overcrowding is avoided and plants do not need to be moved again until planting out. Trays may be stood outdoors.

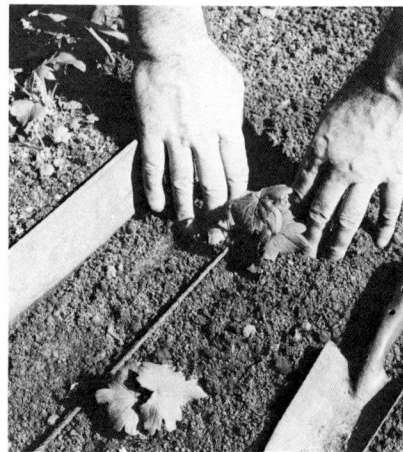

Young plants can also be pricked out into 3½-in pots when a few especially good batches of plants are needed or when planting out may be delayed.

Seedlings which have been grown on in boxes and have become strong plants can be planted into a nursery bed to over-winter. The soil should be rich and fibrous. Set plants out 9 in (23 cm) apart.

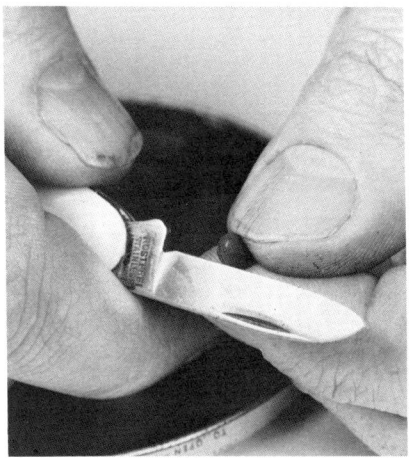

Before sowing prepare a seed label for each variety. Seed bought by name is packed with the variety stamped on the packet, and a seed label should be prepared for each before the packet is opened. Use a waterproof pencil.

Chipping the skin of sweet pea seed. This is a good practice to follow with all hard-skinned seeds as it helps towards even and speedy growth. Use a small penknife to lift a tiny portion of the skin.

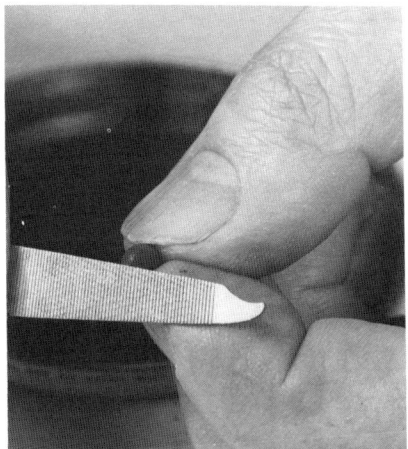

A nail file may also be used to smooth away the skin of a hard-skinned seed until the kernel is exposed. Whether a knife or a nail file is used, the seed should be treated on the opposite side to the eye.

Pots may still be bought for sweet peas, deeper than normal so that the long roots which are made have plenty of space in which to develop. Sow with a dibber, five or six seeds to a pot.

Where a variety of seeds is being grown, 3½-in pots are suitable containers. Once the first signs of germination are seen the pots should be stood where they will benefit from the maximum amount of daylight.

Stand peat or similar pots in a seedtray or shallow box for ease of handling. Place 1 in (2.5 cm) of peat at the bottom of the box. This will help to keep the pots moist.

Seedlings being pricked out (transferred and spaced out) into 2-in (5-cm) paper tubes. Use J.I. potting compost no. 1 or a peat-based equivalent for this purpose. Handle the seedlings by the leaves and set them so that the lower leaves are just above the surface of the compost.

A peat block with cavities filled with J.I. potting compost no. 1 or a peat-based equivalent. The seedlings are pricked out in these and the block stood in a seedtray. On planting out, the block is divided into single plants.

Seedlings being pricked out into a seedtray. Here two different subjects are being grown on in one tray. Flower seedlings can be pricked out fifty-four to a seedtray, i.e. in rows nine by six, but for small batches forty to each tray is adequate.

A good example of spaced sowing of gloxinias. The seeds are spaced out at sowing time and pricking out may not be necessary, particularly if the young plants are to be grown in pots subsequently.

After the seedbed has been dug or forked through, break down the surface soil with the tines of a fork or a cultivator to obtain the necessary tilth (fine soil) in which to sow the seeds. Prepare an especially fine bed for small flower seeds.

A wooden rake is a most useful tool for breaking down the surface soil in preparation for seed sowing but an ordinary rake can be used equally well. Biennials or perennials sown for cut flower purposes, as well as vegetables for the kitchen, all require the same preparation.

A dressing of superphosphate may be applied before sowing at 3 oz per sq yd (90 g per sq m) to ensure good root development for seedlings and young plants. Rake in the fertilizer evenly.

If the soil is heavy, add a half-and-half mixture of peat and sand at the rate of 1 lb per sq yd (500 g per sq m). Rake this into the top few inches of soil.

Seed of many perennials like lupins, delphiniums and scabious may be sown outdoors in May and June. Drills should be $\frac{1}{2}$ in (1 cm) deep and 6 in (15 cm) apart, drawn out with a hoe against a long rule or garden line.

Mix the seeds with sand so you can gauge the rate of sowing. Sow thinly so that the seeds are spaced about $\frac{1}{2}$ in (1 cm) apart. This will prevent overcrowding and ensure better development of the young plants.

Cover the seeds by drawing the soil into the drills to an even depth with a rake. Firm the soil evenly with the head of the rake.

Label each row at the time of sowing. A small seedbed with several short rows will be adequate for average home gardening purposes.

Another method of drawing drills is to use the handle of a rake or hoe, pressing this into finely prepared soil to a depth of approximately ½ in (1 cm).

Drills can also be made with the back of the rake head. An even depth of drill is important, otherwise germination will be erratic.

If the soil is very dry, draw the drills 2 in (5 cm) deep and fill in 1 in (2.5 cm) at the base with moist peat, or water along the drill before sowing.

After sowing cover the drills with a sprinkling of grass mowings as a protection against birds. Seeds rows are easily damaged by birds 'dusting' in dry soil.

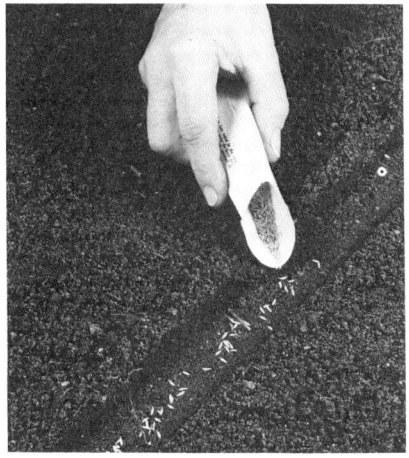

Seeds can be sown in a number of ways. Most are sown thinly along the length of the drill.

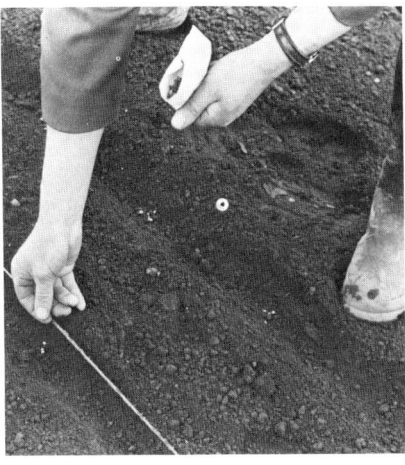

Some seeds, such as beetroot, can be spaced at regular intervals. This is known as station sowing. About three seeds are sown at each station. When germination takes place thin to leave the strongest seedling.

Seeds can be broadcast over a block of ground or a wide strip to give a higher yield. Peas are broadcast in a 6 in (15 cm) deep drill taken out with a draw hoe against a garden line.

Scatter the pea seeds over the base of the drill and cover with 2 or 3 in (5 or 8 cm) of soil using a hoe or a rake. This will leave a trench about 3 in (8 cm) deep which will give protection to the young plants as well as allowing them to be flooded with water in dry periods.

This is a method of pre-germinating seed. It can be used for seeds that are slow to germinate or for those that are to be sown in dry conditions. The success rate can be improved if these seeds are pre-germinated. The seeds are placed on a wad of damp tissues or blotting paper and kept in a warm, dark place until germination occurs.

As soon as the tiny roots are seen, remove the seeds and mix them into some previously prepared Polycell or other wallpaper adhesive or into a gel specially prepared for the job.

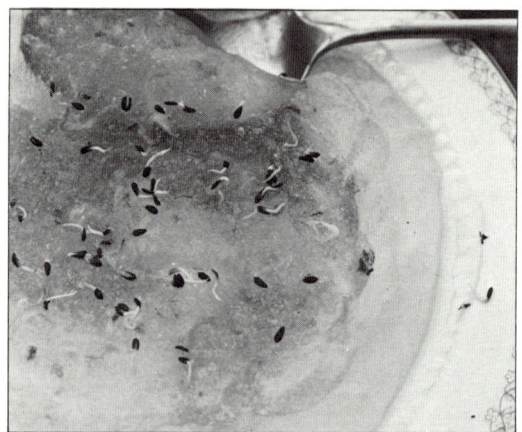

The gel and seed mixture is put into a polythene bag with one corner snipped off. The gel can then be squeezed into a prepared drill. Alternatively use one of the bottles specially manufactured for the purpose. The gel protects the seeds against dry conditions and enables them to be spaced more evenly in the row. This does away with the first thinning which causes so much root disturbance. The seeds and gel must, of course, be covered with soil in the usual way.

Opposite: A brilliant display of mixed annuals grown from seed

Lawn seed can be sown in late August or early September, or in spring. Dig the site about a month before sowing to allow time for the soil to settle. Tread or roll the area when the soil is dry, to break down lumps and to give even firmness.

Use a rake to break down the soil and to obtain the tilth (fine surface) needed for sowing. Several such rakings should be given, but only when the soil is dry and will not adhere to the tines of the rake. Remove any stones whilst raking.

Use a springbok or ordinary rake to obtain the final tilth. A base dressing of fertilizer can be raked into the soil surface at this stage. A balanced general fertilizer such as Growmore would be suitable for the job.

Roll to obtain the final level and then give a further raking. Aim at an even surface with no bumps or hollows, and an even tilth over all the plot.

Opposite: Mountain ash can be grown from berries stratified before sowing. Choice varieties are budded onto a *Sorbus aucuparia* rootstock _____ 35

For large areas, a spreader may be used for seed sowing. Seed today is, in most cases, treated against birds, but shake the packet well before filling the machine to ensure that the seed is well coated with the bird deterrent. Load the hopper with the required amount of seed. The rate of distribution is controlled by changeable roller bars.

Sowing in progress. With the spreader even distribution is assured, and large areas can be dealt with quickly and evenly. Sowing rate advised is 1½ to 2 oz per sq yd (60 g per sq m).

After sowing, lightly rake the whole area with a springbok rake. Seed should be only just covered.

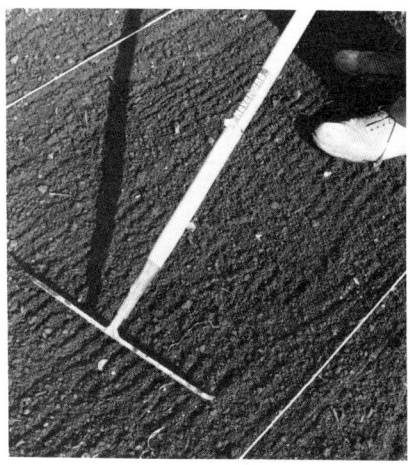

To simplify sowing a lawn by hand the plot is best marked off into yard- (metre-) wide strips with string. These strips can then be measured off to find the total number of square yards (metres).

Before sowing each strip, work from one end backwards, raking the ground down to an even level and tilth. Ensure that all footmarks are removed.

Having marked out the number of square yards (metres) in each strip and determined the rate of sowing, usually 2 oz per sq yd (60 g per sq m), place the seed in a bowl. Take a handful and scatter the seed evenly, keeping the hand close to the ground.

Most of the good seed offered today is treated against its biggest enemy – the birds. But it is still a help to germination to cover it in by lightly raking over the ground. Both sowing and raking should be done standing on a board at the side of the strip.

Cuttings

A cutting is a piece of a plant – shoot, stem, leaf, bud or root – detached from its parent and encouraged to produce a new plant or plants. These new plants will be exact replicas of the parent, unlike many of those raised from seed. The latter can give rise to considerable variation during the exchange of genetic material which occurs with pollination. A cutting although incomplete in itself is capable of generating those parts of the plant initially absent but essential for growth. As the only genetic material present in the cutting is that of the one parent, these new parts when they grow will resemble those of the parent in every way. This is, therefore, the way named varieties of flowers and shrubs are increased, although, of course, species and natural varieties can be reproduced in this way too.

Perhaps the type of cutting that first comes to mind is the stem cutting. This consists of a stem with buds and can include the shoot tip, although this may be removed to induce side branching. Most of the leaves can be left on as is usual with softwood cuttings or they can all be removed as in the case of many hardwood cuttings. Softwood cuttings are, as the name suggests, taken when the stem is still soft and sappy in spring or early summer. Half-ripe cuttings are taken in summer when the shoots are firmer and can be rooted in slightly cooler conditions. Hardwood cuttings are taken in the late autumn or early winter when the stem is hard and no longer green. These are then planted in open ground for lifting the following year.

Many herbaceous plants are propagated from stem and shoot cuttings. Chrysanthemums, dahlias and lupins, to name just three examples, are cut back in the autumn and thus encouraged to produce new growth in spring or even in winter if forced with some heat. This fresh young growth provides shoots that can successfully develop into new plants if treated correctly.

Some plants produce leaves which are capable of generating new plants if they are cut into pieces or scored across the veins. Streptocarpus, sansevieria and *Begonia rex* are the three that are most commonly propagated in this way, although the variegated sansevieria must be reproduced by division or the golden edge to the foliage will be lost. Sedum, Christmas cactus, peperomia and saintpaulia are four plants that will produce a cluster of new plants at their leaf bases if the leaves are inserted into a sandy compost. Leaf bud cuttings need the bud in the leaf axil to be present in order to produce a new plant. Camellias, ficus (rubber plant) and pelargoniums can be propagated in this way, although camellias can take up to a year to show signs of shooting.

Pure bud cuttings, known as eyes, are taken in late autumn or winter when all the leaves have fallen. The plants most commonly reproduced in this way are wisteria and grape vines. It is an economical method of propagation as one plant can yield a very large number of cuttings.

Root cuttings consist of pieces of root alone; none of the above ground parts of the plant are necessary at all. They are often known as thongs. These thong-like roots are cut into pieces and must always be planted the right way up. Seakale, romneya, perennial poppies and eryngium are just some of the plants that can be increased by this method. Some plants like mint, phlox and raspberries produce buds on their roots. If little pieces of root are sown, almost as seed, the bud will shoot and produce a new plant.

These are the main types of cuttings that can be taken, but in order to ensure success various other points must be followed. Good hygiene is all important. Cuttings should only be taken from strong, healthy plants. The compost should be fresh and of the right composition. Garden soil is not suitable. A gritty compost containing plenty of sharp sand (never builder's sand) is generally advisable. A mixture of one part sharp sand to one part peat suits most plants but special rooting composts are available. As soon as the cutting has produced roots and is beginning to grow it should be transplanted into a stronger medium containing more fertilizer and trace elements, such as John Innes potting compost no. 1 or no. 2 or a peat-based equivalent. Cuttings of calcifuges should never be inserted in a compost with the slightest trace of lime. Special ericaceous composts are available for camellias and lime-hating heathers.

Secateurs and knives used to take cuttings should be clean and sharp. Disease can be transferred on a knife blade and a jagged cut can provide an entry point for damaging organisms. All seedtrays and pots should be well scrubbed out and rinsed clean. Free drainage and adequate ventilation are all important otherwise there can be problems with damping off fungus (see page 12). Care is needed when inserting a cutting to ensure the base is in contact with the compost. If a dibber is being used make sure it is the correct size and not too large or the cutting will sit mostly in air spaces.

There are a number of aids which will help cuttings to root successfully. Hormone rooting powder is often recommended especially for shy-rooting subjects. It is only necessary to dip the base of the cutting into the powder and then to tap off the excess; too generous a coating is of no benefit. A propagating case or a frame

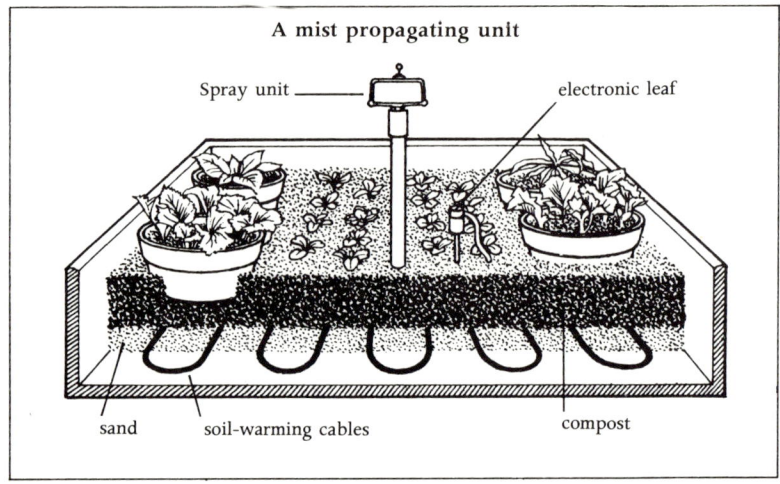

A mist propagating unit

Spray unit

electronic leaf

sand   soil-warming cables   compost

fitted with soil-warming cables to provide bottom heat will help to speed up rooting and can be used in conjunction with a mist propagation unit. This sends out a mist of water over the foliage at regular intervals which cuts down transpiration and enables the cuttings to put all their energies into forming roots. Mist propagation is especially useful for evergreens and other slow-rooting plants. A simpler and cheaper method when dealing with just three or four cuttings, or even less, is to arrange a polythene bag over the pot. Check the polythene is not in contact with the foliage and hold it firmly in position with an elastic band around the rim of the pot. When the cuttings show signs of growth remove the bag and pot on the cuttings in the normal way.

Softwood cuttings of shrubs use young, short, soft side shoots 3 to 4 in (8 to 10 cm) long taken in late spring. Prepare by removing the leaves from the lower half of the cutting.

Cut straight through at the base just below a leaf joint, using a sharp knife or razor blade.

Dip the bases of the prepared cuttings in hormone rooting powder before insertion. This treatment gives quicker and better rooting and is especially valuable for those subjects that root less readily.

Insert cuttings in a cold frame in a sandy compost at 3-in (8-cm) spacing. The frame should be kept 'close', i.e. very little ventilation should be given, until the cuttings are rooted. Alternatively they can be inserted in pots which are either placed in a propagating case or covered with a polythene bag.

The old stools are planted in boxes in the autumn and then kept cool and fairly dry until January when they are brought into warmth and watered. The cuttings are taken when growth is well developed; shoots 3 in (8 cm) long and of diary pencil thickness are suitable. Do not use thick or overgrown cuttings.

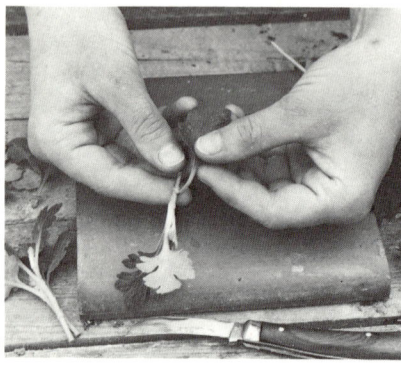

To prepare cuttings, first remove the leaves by pulling them away from the stem. The growing points are extremely tender and brittle, and care in handling is essential.

Next prepare the base by cutting through the shoot to leave it 2 to 2½ in (5 to 6.5 cm) long. Cut cleanly just below a leaf joint with a single-edged razor blade.

Dip the base of the cutting into hormone rooting powder and tap off the excess. This aid is especially useful if only a few cuttings are being taken and hundred-per-cent rooting is required.

For rooting small batches of cuttings $3\frac{1}{2}$-in pots are best. Cover the compost in each pot with a layer of sharp sand; a little of this will trickle into the holes and aid rooting.

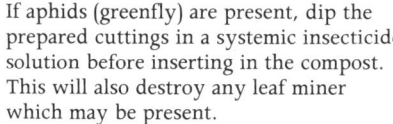

If aphids (greenfly) are present, dip the prepared cuttings in a systemic insecticide solution before inserting in the compost. This will also destroy any leaf miner which may be present.

Cuttings root more readily when inserted round the edge of the pot. Six is an ideal number for the size of pot shown here. Each cutting must be made firm.

When space is limited, strike cuttings in seedtrays, keeping varieties in rows. Label each row for identification.

Dahlia tubers are usually lifted at the end of October, after the first frost, and stored in dry peat over winter. Start these tubers into growth in a heated greenhouse in February. Select firm growths for cuttings and do not use any that are thick or hollow.

Prepare the cuttings by removing the lower leaves. Shoots may be taken with a small piece of parent tuber still attached; in this case it is not necessary to trim below a joint.

Use a razor blade of the type shown here to prepare the cuttings where a piece of tuber is not attached. Cut cleanly just below a leaf joint, using a firm surface.

Dip the base of each cutting in hormone rooting powder. Tap to remove surplus powder.

Small batches of cuttings are best planted in 5-in pots. Spread a thin layer of sharp sand over the surface of the compost and make it firm and level.

Ensure that the base of each cutting is touching the compost at the bottom of the hole when it is inserted. Firm the compost around each cutting as it is put in. Take care not to overwater as the cuttings may rot if kept too damp.

After 2 or 3 weeks, spread the fingers through the cuttings, turn the pot over and carefully knock the plants out. Roots should be showing and the cuttings ready for potting on singly in $3\frac{1}{2}$-in pots and J.I. potting compost no. 1 or a soilless mix.

When potting handle the plants with care, as the roots are brittle. Do not plant out of doors until all danger of frost has passed.

Take young shoots for use as cuttings when 3 to 4 in (8 to 10 cm) long in spring. To obtain earlier cuttings, lift a few crowns and plant in a cold frame in winter.

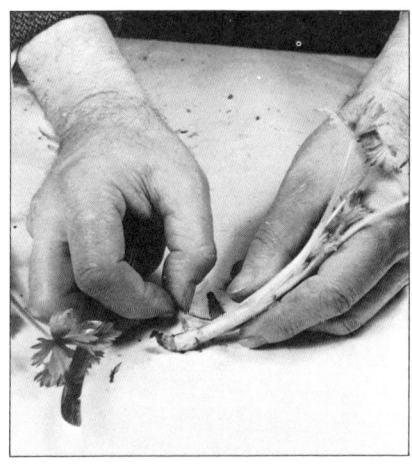

Strip off the lower skin from the base of the shoot with the fingers. No leaves should be removed from young cuttings.

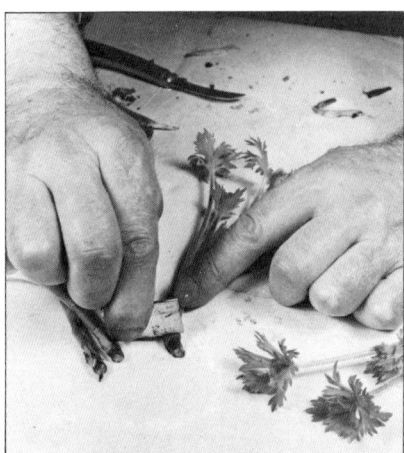

Cut through the base of the shoot cleanly with either a sharp knife or a razor blade.

Insert the cuttings in a sandy compost in pots, setting them firmly round the edge. Do not overwater. Keep the pots in a cold frame or cold greenhouse until the cuttings are well rooted, then plant outdoors.

Take young shoots, produced by plants stored in the greenhouse, early in spring as 2 to 3 in (5 to 8 cm) cuttings. Remove the two lower leaves and cut cleanly through below the leaf joint with a sharp razor blade.

Plant the cuttings fairly close together in 5-in pots. A suitable rooting medium, in which the cuttings can also grow on for a while, is three parts loam, two parts sharp sand and one part peat (by bulk).

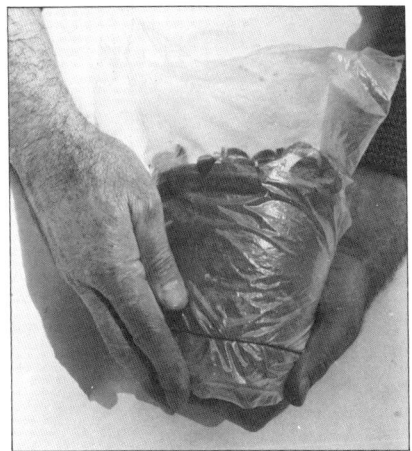

Cover each pot with a polythene bag. This encourages quicker rooting and helps to stop wilting. Leave 4 in (10 cm) between the top of the cuttings and the bag. A rubber band round the bottom of the pot holds the polythene in place.

Strong-rooted plants will develop quickly with a little care. If plants are intended to be grown on for ornamental use in greenhouse or home pot them on into 5-in pots. If the plants are for bedding, 3½-in pots will be sufficient.

Take a shoot from low down on the side of the parent plant in spring when the plants are being grown in the greenhouse or in summer (June or July) if the plant is growing outside.

Cut through the base of the shoot between two buds (i.e. at an internode) leaving the tip intact. Cut off the lower leaves cleanly.

Insert the cuttings firmly in vermiculite or a sandy compost such as three parts sharp sand and one part each peat and loam. Water with care and never over water.

Rooted cutting ready for potting on singly into a 3-in pot. If blue flowers are required, do not use a potting compost containing lime at any stage. Blueing agents, such as sequestrene or aluminium sulphate, can be applied annually to retain the blue colour. Pink blooms are enhanced by a dressing of lime.

Where further plants are required from a selected plant, such as one with a particularly fine colour, young shoots may be taken in spring when 4 or 5 in (10 or 13 cm) long for use as cuttings.

A selection of shoots taken from the parent plant. Avoid using very thick, very thin or weak shoots.

Prepare the shoots by trimming off the lower leaves and cutting cleanly through the base with a sharp knife or razor blade.

Insert the cuttings round the edge of a 5-in pot or half pot, using a compost of 3 parts loam, 2 parts sharp sand and 1 part peat with a thin layer of sand on top. Keep in a cold frame or a cold greenhouse until cuttings are well rooted.

A cutting prepared for insertion. The basal cut is made just below a leaf joint and the lower leaves removed cleanly. A rooting hormone powder applied to the base speeds rooting.

The cuttings are inserted round the edge of a 5-in pot. A sharp sandy compost is best: 3 parts sharp sand to 1 part each peat and loam. Ensure that the base of each cutting is touching the compost and make very firm.

Deep seed boxes may be used instead of pots where space is limited. Make up the boxes with the same compost as for pots and plant the cuttings 3 in (8 cm) apart. Water the soil only, not the leaves, to prevent damping off.

When cuttings have been taken in the autumn and become strong-rooted plants, the tips can be cut off in spring and used as further cuttings.

Opposite: Both achillea and rudbeckia can be propagated from seed or by division

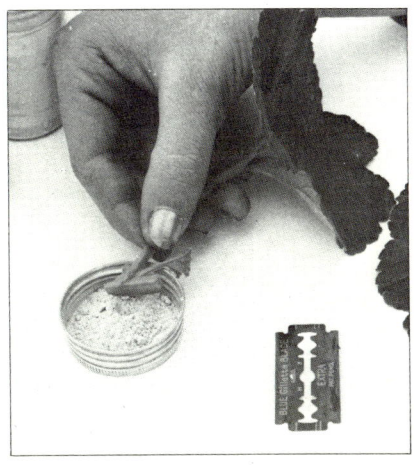

Pieces 1-in (2.5-cm) long containing a leaf joint, with a leaf bud showing, make an economical method of propagation. Cut ½ in (1 cm) above and below the joint.

Cut the 1-in (2.5-cm) stub in half through the middle of the stem; discard the half without the bud and dip the half with the bud into the hormone rooting powder, ensuring that the surface is well covered. Tap off any excess.

Pot each cutting separately, making sure that only the top is left above the soil. Firm evenly all round with a dibber. A peat pot which can be kept damp by watering from the bottom is best.

A young rooted bud cutting ready for potting on. These cuttings make roots fairly quickly and soon fill the small peat pots.

Opposite: Magnificent delphiniums like these can be raised from cuttings or seed _____ 53

Suitable cuttings can usually be found at the ends of the main and side shoots. Take cuttings about 6 in (15 cm) long with a sharp knife. Prepare them by removing the lower leaves and cutting through the stem just below a leaf joint so the cutting is about 4 in (10 cm) long.

Prepare a 5-in pot with a sand, peat and compost mixture (in equal proportions) and cover the surface with $\frac{1}{4}$ in (0.5 cm) of silver sand, so that a little of the sand will go into each of the holes as the cuttings are inserted.

The medium used for striking cuttings does not contain sufficient nutriment for the plants to grow on once they are rooted. Cuttings should therefore be potted on as soon as possible after a good root system has been established.

Suitable cuttings can be found around the outsides of the old clumps of pinks and border carnations, and are best taken in June or July. Remove lower leaves by pulling them off.

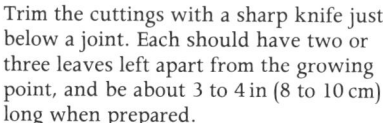

Trim the cuttings with a sharp knife just below a joint. Each should have two or three leaves left apart from the growing point, and be about 3 to 4 in (8 to 10 cm) long when prepared.

Although pinks and carnations strike quite easily, a dip in hormone rooting powder guarantees quicker rooting. It is simplest to bunch a number of cuttings together when doing this and then shake off any excess powder.

Cover the compost in the pot with silver sand and insert the cuttings around the edge of the pot to about a third of their length and 1 in (2.5 cm) apart, ensuring that each is quite firm. Give a good watering using a can with a fine rose.

This is a method of rooting a number of plants that can be propagated by softwood cuttings. Take a 12-by-6-in (30-by-15-cm) strip of polythene and lay a $\frac{1}{4}$-in (0.5-cm) layer of damp moss along half the width.

Prepare cuttings by the normal method. Dip them in hormone powder and lay them side by side on the moss $\frac{1}{2}$ to 1 in (1 to 2.5 cm) apart.

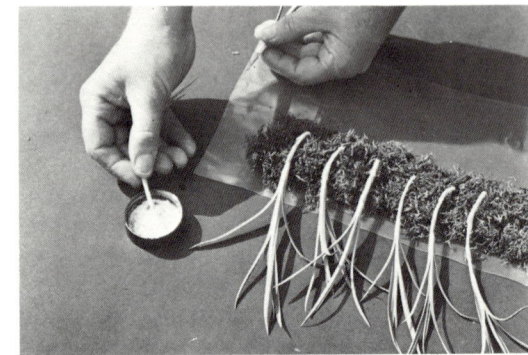

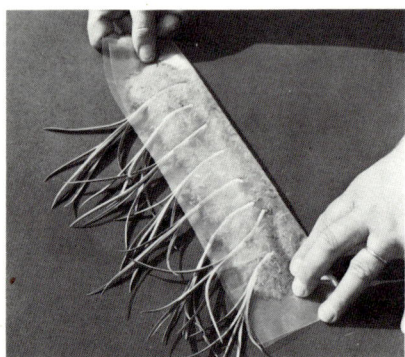

Fold over the other half of the polythene to cover the base of the cuttings.

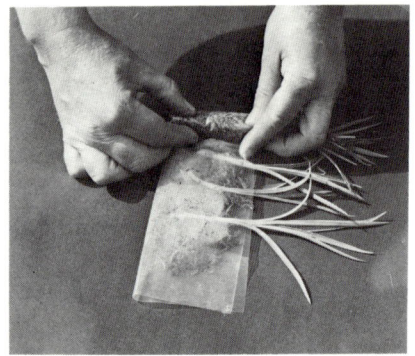

Then roll up the cuttings in the polythene, keeping the moss on the outside. Keep the roll firm and secure with a rubber band. Stand in a warm, light position and inspect regularly. As soon as roots develop, plant into a seed box containing sandy loam.

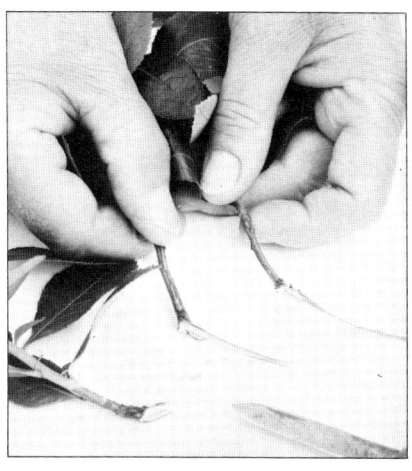

A heel cutting is a type of half-ripe cutting (see overleaf). Pull off a side-shoot 3 or 4 in (8 or 10 cm) long with a 'heel'. Hold the shoot close to the stem with finger and thumb and pull downwards towards the base. June or July is the time for taking half-ripe cuttings.

A length of the parent stem will be attached to the heel shoot: pare this with a sharp knife.

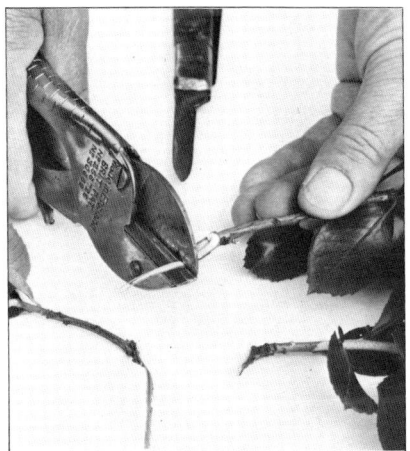

Trim the long tail on the heel, leaving about ½ in (1 cm) of old wood. Cut cleanly through with secateurs.

Trim rough edges cleanly with a knife and remove lower leaves.

Half-ripe cuttings of shrubs are taken in June and July when the stems are soft. Cut them about 4 to 6 in (10 to 15 cm) long and trim off the lower leaves.

If a number of different plants are to be propagated it is a good idea to insert them in a bottomless box placed in a sheltered part of the garden out of the sun. See the underlying soil has been worked to a fine tilth; a little sharp sand can be incorporated with it if it is heavy. Dip the base of each cutting in hormone rooting powder, tapping off the excess.

Plant each cutting firmly in rows across the box according to species. The rows should be about 4 in (10 cm) apart.

When the cuttings are all inserted and the box is full, water them in using a fine rose on the watering can.

Cover the box with a sheet of glass to cut down on the water loss from the cuttings which can be considerable during the summer months.

Inspect the cuttings regularly and as soon as they show signs of growth remove the glass. After about ten weeks a good root system should have developed and the plants can be moved to grow on with more space round them.

Take off side-growths from the base of the plant in June. It is always best to replace lavender bushes before they become straggly.

Many half-ripe cuttings, including lavender, root better if a small heel of the old wood is left.

Dip the cuttings with lower leaves removed in a hormone rooting powder before insertion.

Insert the cuttings in a seed tray, in a compost of three parts sand, two parts loam and one part peat, at 2-in (5-cm) spacing.

Half-ripe cuttings of conifers, in this instance juniper, are taken by pulling small side shoots away from the stem, each of which will have a heel of old wood still attached.

Dip the base of the cutting in hormone rooting powder. This will help speed up the production of roots.

Conifer cuttings may be difficult to root and are, therefore, ideal subjects for propagating under a mist spray. They can be inserted into a rooting medium of peat and sharp sand laid directly on the bench or potted up first. The mist spray cuts down on natural water loss from the cuttings and speeds up rooting.

A well-rooted cutting after it has been lifted from the bench. It should now be potted up in a 3-in pot in John Innes potting compost no. 1 with some extra peat or a soilless equivalent. For preference, obtain an ericaceous compost specially formulated for acid-loving plants.

Cuttings can be taken in autumn from a wide variety of shrubs. Select young shoots made during the current season. Prepare cuttings of privet from shoots about 24 in (60 cm) long. Discard the thin end and cut just below a bud at the base so the cutting is 10 in (25 cm) in length. The same preparation is suitable for the other hardwood shrubs.

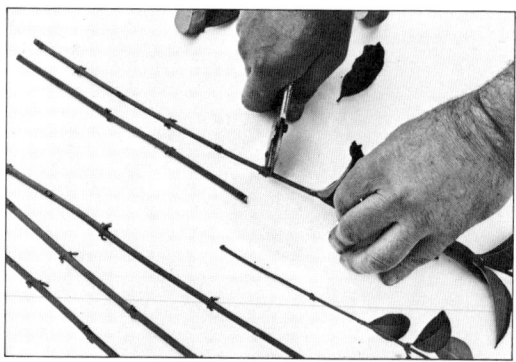

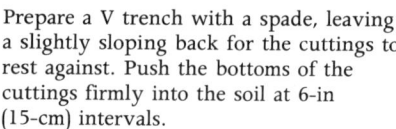

Prepare a V trench with a spade, leaving a slightly sloping back for the cuttings to rest against. Push the bottoms of the cuttings firmly into the soil at 6-in (15-cm) intervals.

Firm the cuttings in well after insertion. It may be necessary to repeat this after a severe frost. Signs of shoot growth in spring will indicate that roots have formed. The cuttings can be moved to their permanent quarters in the autumn.

Alternatively the cuttings can be bundled and labelled according to variety. Dip the bases into hormone rooting powder and heel in in a protected area of the garden or in a cold frame. Lift in early summer, by which time roots should have formed.

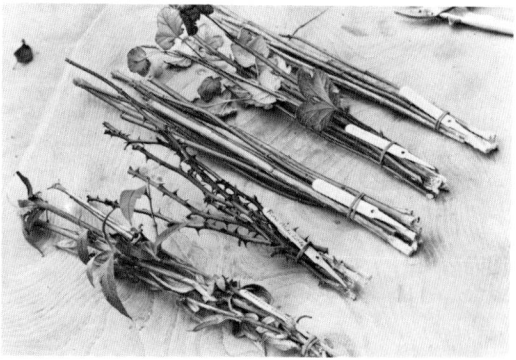

Hardwood cuttings of evergreens such as *Senecio laxifolius* can be taken in autumn in a similar manner to deciduous plants. Take cuttings of firm wood about 8 to 10 in (20 to 25 cm) long, remove the lower leaves and dip the bases in hormone rooting powder.

Plant out in a V-shaped trench about 6 in (15 cm) deep. A layer of sharp sand at the bottom of the trench will encourage rooting to take place.

Fill the trench with soil and firm the cuttings well into place with the feet.

It is a good idea to allocate a corner of the garden as a nursery bed. The cuttings can then be left undisturbed until they show signs of shooting, with just an occasional hoe between the rows to keep weeds at bay.

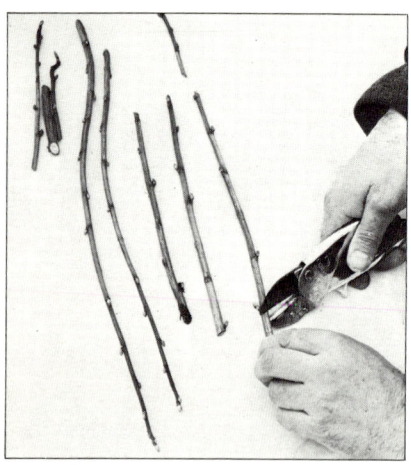

Prepare cuttings 8 to 10 in (20 to 25 cm) long in autumn, using the new shoots made that season. Cut above a bud at the top, below one at the base. Do not use the thin ends of the shoots.

Using a line, make a slit trench with a spade. The slit should be 6 to 7 in (15 to 18 cm) deep.

Insert the cuttings in the slit 1 ft (30 cm) apart so that only about 2 in (5 cm) remain above soil level. Note that all the buds are left intact on the cuttings.

Make the cuttings firm by treading alongside and between them.

Select young shoots of summer growth for cuttings. These are best taken in early autumn.

Prepare cuttings 10 to 12 in (25 to 30 cm) long and discard the thin (upper) ends of the shoots. Cut just above a bud at the top and just below a bud at the bottom.

Remove the lower leaves and, if the bush is to be grown on a leg, all but the top four buds. If, however, the bush is to be trained some other way, leave all the buds intact.

Insert the cuttings 1 ft (30 cm) apart in a trench 6 in (15 cm) deep, leaving about a third of the cutting above soil level. Firm planting is essential.

Leaf-bud cuttings of camellia
should be taken in March.
Make sure each leaf has a good
bud in its axil. Take the
cutting with a sharp knife and
remove the sliver of bark
behind each bud and leaf.

Insert three leaf-bud cuttings
round the edge of a 3½-in pot
filled with a sandy compost
which is completely free of
lime.

The cutting will have taken
when the bud shows signs of
growth. This may take nine
months or more but can be
speeded up by placing the pot
in a frame or propagating case
which has bottom heat. When
roots have formed each
cutting can be potted up
singly in 3-in pots.

The thick, rather fleshy leaves of *Begonia rex* can be cut into 1-in (2.5-cm) squares and laid, underside downwards, on compost in a seedtray. Keep shaded from direct sunlight.

Roots are formed at the edges of the leaf squares and later young plants develop, as seen here. Not all the leaf cuttings develop equally, and some are ready to pot on before others.

An alternative method is to partially cut the thick rib-like veins on the underside of the leaf with a sharp knife at 2-in (5-cm) intervals.

The leaf is pegged down flat on the surface of a pot or seedtray with wire hair pins. Ensure that the cut ribs are in contact with the compost. In time a new plant should form from each cut.

The large, well-developed leaves of this plant are suitable for propagation, and will root fairly readily in a greenhouse or other warm place in spring or summer.

Cut off the leaves and cut the stalk through cleanly with a sharp knife so that about 1 in (2.5 cm) of stem remains.

Insert the cutting firmly in a small pot so that the underside of the leaf rests on the surface of the compost.

Three leaf cuttings can be inserted in a 3½-in pot. A suitable compost is made up of half peat and half sand.

Remove single leaves from
saintpaulia (African violet)
plants that are growing well
and have good healthy foliage.
Trim the leaf stalk so that it
measures about 1 in (2.5 cm).

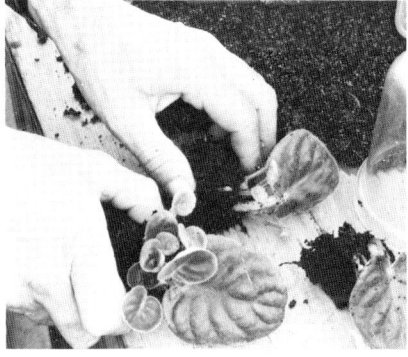

Insert four or five leaf cuttings round the
edge of a 3½-in pot containing a soilless
compost. Put this in a warm place, if
possible with bottom heat, and keep the
compost moist.

Each leaf should produce several shoots.
When this happens remove the cutting
from the pot and it will be apparent that
each leaf has produced a number of tiny
plants. These should be gently eased
apart.

Pot each tiny plant up
separately in a 3-in pot in J.I.
potting compost no. 1 or a
peat-based equivalent.

Detach a leaf from a mature and healthy streptocarpus plant and lay it down flat on a seedtray full of compost topped with a layer of sharp sand. Pin it down so it is in contact with the sand and slash the main vein two or three times.

In time new plants will form where the main vein was cut if the compost is kept moist.

These new young plants can be carefully lifted and potted up in 3-in pots filled with John Innes potting compost no. 1 or a peat-based equivalent.

When potting up a plant make sure it is well firmed in and there are no air pockets left in the compost. The roots should be in close contact with the rooting medium so that the plant gets away to a good start.

The swollen leaves of the succulent sedums can be broken off and pushed gently into a pan or pot of gritty compost. Keep the compost damp and a new baby plant should be formed at each leaf base.

Detach small pieces of Christmas cactus from a mature plant. The parent plant should have finished flowering and have been fed to ensure it is the peak of health and the foliage is firm and green.

Insert four cuttings round the edge of a 3½-in pot filled with a gritty compost which should be kept just moist. On no account over water.

This is done in a heated greenhouse in late winter whilst the buds are still dormant. Slice the buds or eyes away from the parent shoot with 1 in (2.5 cm) of the shoot attached and then cut away half the thickness of the parent shoot.

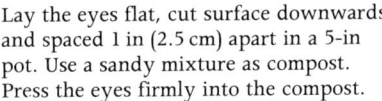

Lay the eyes flat, cut surface downwards and spaced 1 in (2.5 cm) apart in a 5-in pot. Use a sandy mixture as compost. Press the eyes firmly into the compost.

Cover the surface of the pot with a $\frac{1}{4}$-in (0.5-cm) layer of sharp sand after the eyes have been laid in position. Water with care and stand the pot in a propagating case in a temperature of 19° to 21°C (66° to 70°F).

An alternative method of eye propagation is to retain the full thickness of the shoot, but to cut to a length of about $1\frac{1}{2}$ in (4 cm) and insert the eyes firmly in an upright position. The same rooting conditions as mentioned above are necessary.

Mature specimens of dracaena will produce a thick fleshy type of root known as a 'toe' which can be used for propagation purposes.

Loosen the soil round the toe. Then cut the toe off cleanly with a sharp knife, making sure you remember which is the uppermost end.

Making sure the toe is the right way up, dibble it into a pot or pan of compost. Ensure that the whole of the base is in contact with the compost and the top is level with the surface.

When the new growth is established, lift the cutting which will have produced several roots. Pot on into J.I. potting compost no. 1 or a soilless equivalent.

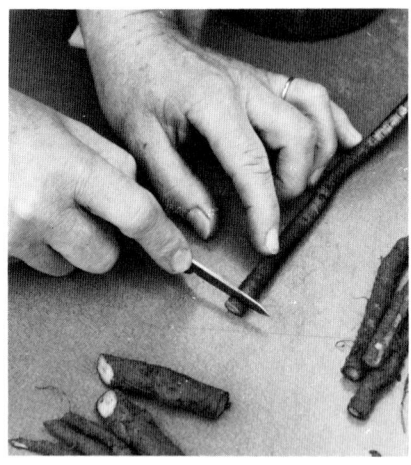

Herbaceous plants with fleshy roots like eryngium, papaver, brunnera and seakale lend themselves to propagation by root cuttings. Lift the parent plant in winter and detach a stout root. The parent crown can be replanted.

Cut straight through the thickest end of the root which was nearest the parent crown. Use a sharp knife and leave a flat surface.

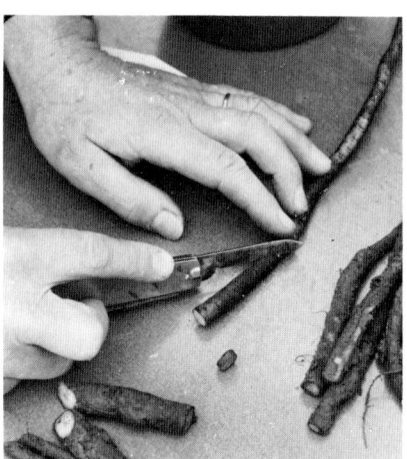

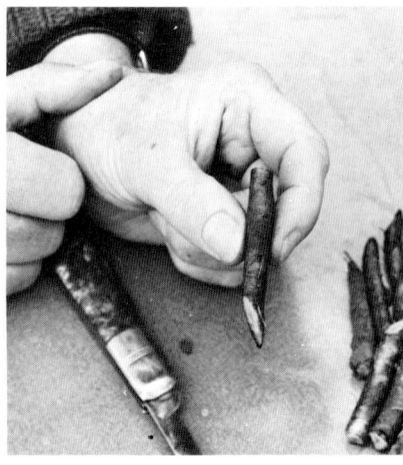

Cut through the other end which was furthest from the crown at a slant. This is to differentiate between the top and bottom of the cutting.

The prepared cutting ready for planting should be about 2 in (5 cm) long. Several such cuttings can be taken from each length of root.

Insert the prepared cuttings in a pot containing a mixture of 3 parts loam, 2 parts peat and 1 part sand. This should be fairly firm and surfaced with ¼-in (0.5-cm) layer of sharp sand. Insert the cuttings so that the slanting end goes in first and the flat (top) surface is uppermost. Tops of the cuttings should be level with the surface of the compost. Insert the cuttings firmly spacing them 1 in (2.5 cm) apart.

Cover the pot with a sheet of glass to encourage quicker rooting, and stand in a cold frame or a cold greenhouse for the winter months.

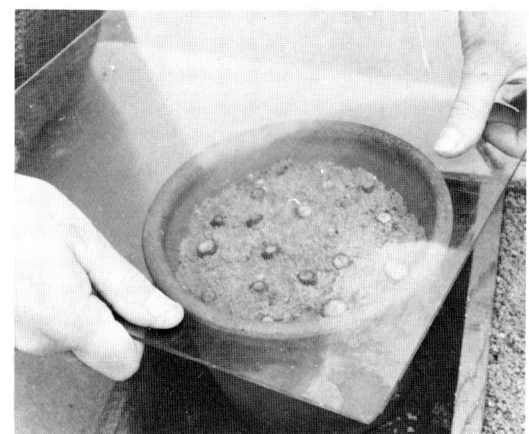

New growths will develop on the top surface of the cuttings. These new young plants can then be planted out in their growing positions in spring.

An alternative method of striking root cuttings of herbaceous plants is in a sand box. Prepare a firm wall of sand in a seed box by means of a wooden batten cut to the same measurement as the inside width of the tray. Lay a row of cuttings against the sand wall.

Prepare another wall of sand to go against the cuttings with the wooden batten.

Cover in the cuttings by pushing the sand firmly into place.

This second wall of sand is now ready for the next row of cuttings. Continue in this way until the box is filled. If cuttings from different plants are put in the same box, take care to label each row carefully.

Opposite: Fuchsia and pelargoniums can be grown from seed or cuttings but *Primula obconica* is only raised from seed

Root cuttings of this type of phlox should be taken when young plants free from stem eelworm are required. (This pest causes a blackening of the lower leaves and severely stunts growth.) Take root cuttings in winter when the plants are dormant.

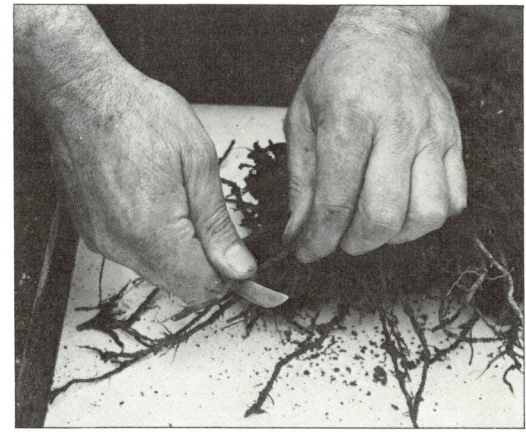

Take cuttings 2 to 3 in (5 to 8 cm) in length and lay them flat on the surface of some sandy compost in a pot or seed tray.

Cover the cuttings with ½ in (1 cm) of sandy compost and place them in a cold frame or cold greenhouse. New shoots will be made in spring, and the young plants can be set in a seed tray to grow on for four to five weeks before being planted out of doors in a reserve plot or nursery bed until autumn, when they can be planted where they are required to flower the following year.

Opposite: Iris pallida Aureo-variegata and other rhizomatous irises are increased by division _____ 79

Division

Division, as its name suggests, involves dividing a plant into pieces. Each piece is a small plant, complete with buds, leaves, stem and roots. This is quite unlike a cutting which consists of only part of the plant and must quickly develop the missing elements if it is to survive. Subjects suitable for dividing must give rise to new plantlets which can be separated easily from the parent plant. In most cases this method requires the parent plant to be lifted from the soil or removed from its pot and its roots freed of soil. However, despite this disturbance to the roots, division usually meets with one hundred per cent success.

A number of herbaceous perennials, such as polyanthus and Michaelmas daisies, form clumps. These can be lifted and eased apart to form either smaller clumps or separate plants. These clump-forming plants can, in time, become very dense. The centre may die out completely in some species and the overall appearance of the plant deteriorates. These plants benefit from being divided every three to five years when any worn-out central portion can be discarded and the outer parts divided and replanted. These, in turn, will need dividing when they have formed a large matted clump.

Rhubarb and paeonies are two examples of plants which develop a fleshy crown just below soil level. These can be cut into pieces to form new plants, provided each piece has at least one bud and some roots present. In a similar manner rhizomes, like those of bearded iris and lily-of-the-valley, and tubers, such as dahlias and begonias, can be cut into pieces. These are both types of underground stem which possess leaves, buds and roots. It follows, therefore, that any section cut to include all these parts should be capable of growing on to form a new plant.

Some plants produce offsets. These are small new plants which grow out from the parent and can be separated and grown on. A number of houseplants such as sansevieria, aechmea and some cacti produce offsets, as do garden plants like globe artichoke and cardoon. Offsets should never be separated from the parent plant until at least a year has passed if success is to be guaranteed. The

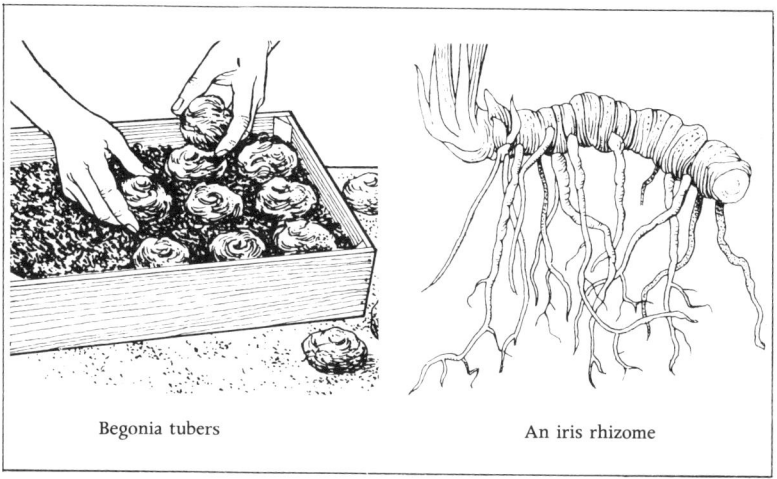

Begonia tubers   An iris rhizome

parent should then continue to grow well and produce further offsets. In addition many bulbs and corms develop offsets but these are dealt with in their own separate chapter (see page 116).

Suckers are very similar to offsets, but the word generally refers to the shoots thrown up from buds on the roots of woody plants. Lilac, _Rhus typhina_ (stag's horn sumach), _Kerria japonica_ and raspberry all produce suckers which can be dug up and separated from the parent plant by cutting through the root close to the base of the shoot with secateurs.

In general, plants should be divided in the spring or autumn and the plantlets should be planted out or potted up immediately. Bulbs and corms require slightly different treatment (see page 116). The usual rules of hygiene apply. Always check the plant is healthy and that the knife or secateurs used are clean and sharp, especially when cutting through rhizomes or crowns. If a division is to be planted in a pot, make sure this has been well scrubbed out and the potting compost used is fresh.

Large tough clumps of herbaceous perennials may be lifted, laid on a firm surface and levered apart by using two forks back-to-back as shown and working the handles backwards and forwards.

Thick tufted clumps which are difficult to lift should be halved and then quartered in situ with a sharp lawn edging tool or spade; the smaller pieces can then be moved without strain. Choose pieces from the outside for replanting and discard the worn-out central portion.

Simple division can be effected by lifting a portion from the side of a plant with a trowel. The worn-out centre of the plant can then be discarded. The best time to do this is early spring when new growth is about to start.

Plants with thick fleshy roots can be divided by cutting through the centre of the crown with a sharp knife. Each portion of top growth must have a section of root.

Lift clumps for division in autumn or spring. Then pull the clumps apart into equal-sized portions.

Each portion so divided can in turn be pulled apart to make further divisions.

Michaelmas daisy crowns of both tall and dwarf varieties can be divided several times.

Each single rooted portion may be planted separately to form a new plant. Best results are obtained by this method of division, which can be done every other year.

Lift the clump carefully with all the roots intact, taking care not to damage the 'eyes'. Divide in early spring or autumn.

Very large clumps should be cut into two equal pieces with a sharp edging tool or a spade with a sharp blade.

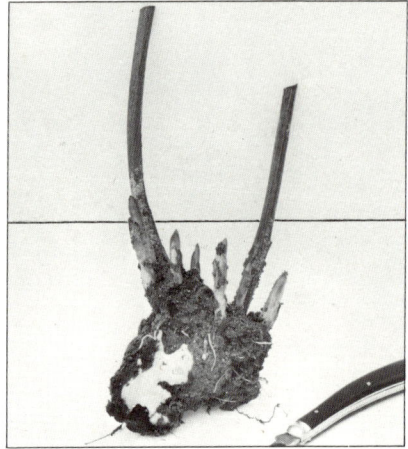

Where the divided pieces are each 6 to 8 in (15 to 20 cm) or more across, they may be divided again. Paeonies do not take kindly to being disturbed and should be divided only when the crowns are very large.

When planting paeonies, do not cover the tops of the crowns with more than 2 in (5 cm) of soil, otherwise flowering may be delayed. The line indicates the correct soil level.

Division of the bearded (June-flowering) iris may be done in March, but more usually after flowering in August or early September. Divide overcrowded clumps to improve the quality of the flowers. A rhizome, lifted and ready for division, is shown here.

Cut off separate portions of the rhizome from the parent clump. Use the outer, younger portions and discard any soft and old pieces.

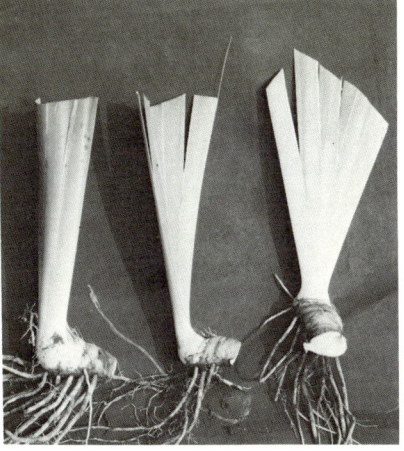

Each separated portion should have roots attached. Division can be continued down to single pieces, so long as each has some root.

Cut back tops of the leaves and trim basal portions of the rhizome neatly with a sharp knife. Replant so that the top of the rhizome remains exposed above soil level.

Lift clumps of polyanthus primulas after flowering in spring and shake off the soil so that the root mass is exposed. Division is best done in dull weather when the soil is damp.

Pull the clump into pieces using gentle pressure of the fingers. Polyanthus lends itself readily to division, each parent plant giving from six to eight pieces.

A clump of *Sedum spectabile* of a size which can be readily divided. Many herbaceous plants give better quality flowers if divided every two or three years.

A clump of geum that has just been lifted. Divide this into two or more pieces by cutting the rather woody crown with a sharp knife. After cutting, pull the portions apart to obtain the young outside pieces which should be replanted as soon as possible.

To increase your stock of a favourite variety of tuberous begonia, cut the tubers into two portions when growth has started and the new shoots are well developed in early spring. Cut the tubers cleanly in half to leave a new shoot and some root on each portion.

It is essential for further development that roots are left on each cut portion. Handle the cut tubers with care so that the roots are not broken or damaged.

Pot each portion into a 3½-in pot, using a compost of half peat and half John Innes potting compost no. 1. Do not pot too firmly. The top of the tuber should be level with the surface of the compost. Once some top growth has been made pot on to a 5-in pot later, using just John Innes potting compost no. 1.

Offsets develop from the base
of mature sansevieria plants.
After about a year these can
be separated and grown on.
Knock the plant out of its pot
and tease away the soil. Ease
the offset away from the
parent plant and cut through
the point of attachment with a
knife.

The offset should have
developed a good root system
of its own and can be potted
up in John Innes potting
compost no. 1 or a peat-based
equivalent.

Place a little compost in the
bottom of the pot and hold the
offset in position. Add more
around the plant, tapping the
pot to settle the compost as
you go, and firming with your
fingers. Water using a can
fitted with a fine rose.

Many shrubs and trees produce shoots from below ground level which are known as suckers. These can be produced close to the base of the parent plant as in hazel. Particular care is needed when lifting the suckers so that the spade does not damage the parent plant.

Other plants such as *Rhus typhina* and lilac throw up suckers over a wider area and so become easier to lift and separate from the parent plant.

All these suckers once lifted can be grown on in a nursery bed or heeled in if the weather is unfavourable to await planting.

When conditions are right the suckers can be moved to their final position in the garden. This is best done in spring or autumn. They will have an established root system and will, in time, produce more suckers of their own.

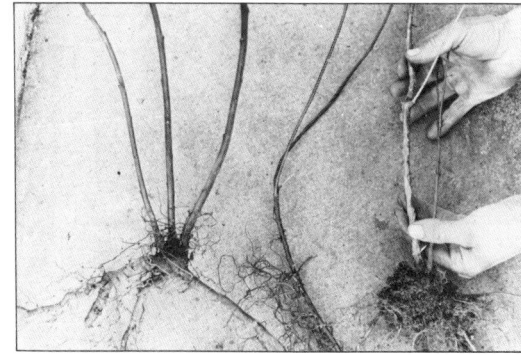

Layering

This is a means of encouraging a stem to produce roots while it is still attached to the parent plant. Some subjects layer themselves naturally. For instance, the tips of blackberry canes will produce roots if they come into contact with the earth. This explains why brambles in the wild form such a thick impenetrable mass so quickly. Other plants produce runners, another natural form of layering, to multiply vegetatively like strawberries, chlorophytum or *Saxifraga stolonifera*. Large woody plants can layer themselves too; if a branch of a honeysuckle or a dogwood comes into contact with the soil roots will eventually develop but, unlike blackberry, the tip carries on growing. This can happen even with large forest trees.

The advantage of layering is that the young plant in its formative stages will still be receiving nutrients from its parent. This means that, in common with divisions, the plantlet once severed can start growing straight away without having to go through the traumatic stage of forming parts, like roots and shoots, which are vital to life. It also means that the new plant is identical in every way to its parent as its genetical makeup is the same. Layering has many points in common with division and therefore there will be some overlap between these two sections.

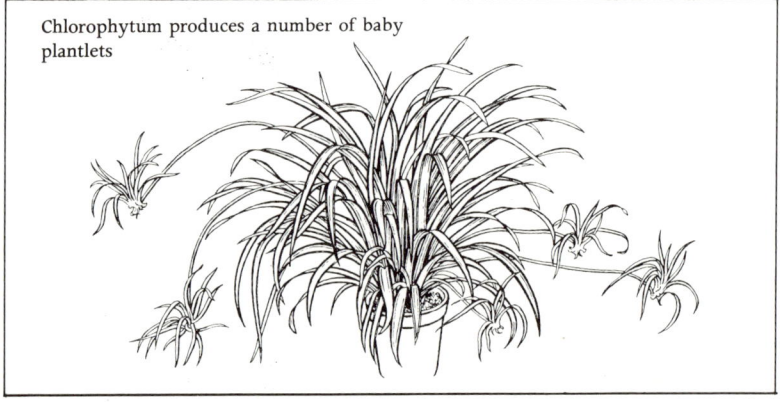

Chlorophytum produces a number of baby plantlets

Layering can be carried out during the spring, summer or early autumn, so long as the soil is moist but not too wet. Choose young flexible shoots that grow low down. These will be encouraged if the shrub was pruned back during the previous season. The stem should be notched or cut in some way where it is to come into contact with the soil. The notch should be held open by either twisting the stem or holding the cut open with a small stone or piece of twig. This will enable the cambium layer to form a callus and it is from this undifferentiated cell growth that the roots will eventually form. It is a good idea to work some sharp sand into the soil surrounding the cut to encourage rooting, especially if the soil happens to be a heavy clay. The layer must be securely pegged down and covered and should be ready for lifting the following spring, although some rhododendrons may take two years or more to root.

Air layering is basically the same operation but the notch or cut is made on a high branch, in fact, in the air, and wrapped in damp moss and sealed in polythene for rooting to take place. This method is used for plants that do not produce flexible or low-growing branches. It is an ideal method for renewing a rubber plant that has lost its lower leaves. The top can be air layered to produce a good-looking new plant. In air layering and other methods where the stem is wounded, hormone rooting powder can be applied to the cut surface to speed up the process and increase the likelihood of success.

Stooling is a form of layering used for propagating black currants and heathers, and for producing rootstocks for roses and fruit trees. Both black currants and heathers can be increased successfully from cuttings but stooling produces new plants in a shorter space of time. It involves bending the stems down into the soil and outwards, exposing the centre of the plant. The centre is then covered with soil which is well firmed down. In time each stem should develop roots where it was bent. In this way a large number of new plants can be raised in quite a short time.

Use shoots on the outside of the clump for layering. Strip off the lower leaves on the shoots to be used. Several shoots can be layered from the same parent clump.

With a sharp knife make a slanting downward cut about 1 in (2.5 cm) in length into the underside of the selected shoot, to about the centre of the shoot.

Peg down the base of the layered shoot into a sandy compost spread on the top of the soil, using a bent piece of wire.

Cover the layered shoots with about a 2 in (5 cm) covering of sand or sandy compost. Layer in July or early August and detach rooted layers from the parent plant the following spring.

In spring or summer select a pliable young branch that can be easily bent down to ground level for layering and make a small cut about 1 ft (30 cm) from the tip. Give the stem a slight twist to open the cut. Dig out a small depression, about 6 in (15 cm) deep, under the cut area of the branch.

Peg the branch down in the depression so the wounded area is in contact with the soil or compost. A bent piece of wire will do the job successfully.

Cover the wounded area of branch with soil and firm down. The leafy tip of the layer can be tied to a stake with raffia to encourage it to grow upright. Water the layer during dry spells.

Rhododendrons can take up to two years to root and should be lifted in spring. When a good root system has formed the layer can be separated from the parent plant with a pair of secateurs.

Clematis can be layered into a pot which can be sunk into the ground. Select a low-growing shoot on the outside of the plant which can be laid down near soil level. Layering should be carried out in June or early July.

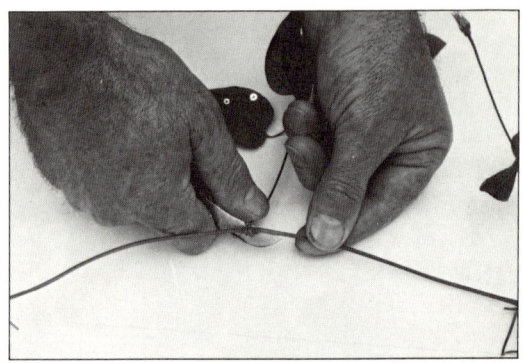

Cut a slanting slit in the stem below a leaf joint. Set a 5-in pot filled with a compost of three parts sharp sand to one part each of loam and peat. Place the pot just beneath the point of cutting.

Secure the portion bearing the cut with pieces of bent wire or hairpins, so that it is set firmly in contact with the compost in the pot.

Cover the layer with compost. Roots will form at the portion of the stem which is partially cut and the rooted layers can be severed in spring. More than one shoot can be layered from the same plant.

Opposite: Layering rhododendrons can take time but is a very effective method. They can also be grafted

This is a method by which some woody subjects can be made to produce roots on high branches. Remove a few leaves and a strip of bark to expose the cambium layer which can be dusted with hormone rooting powder. Then slide a polythene sleeve over the selected shoot.

Position the polythene sleeve over the wound and tie the base firmly in place.

Pack damp moss all round the stem inside the polythene bag, ensuring that the stem is in the centre of the bundle of moss.

Tie the top end of the polythene, so forming an airtight bag. Leave this covering in place until the roots can be seen amongst the moss. Then remove the bag, sever the stem below the roots and plant the rooted part carefully.

Opposite: Commercially-grown bush roses are usually budded onto a *Rosa canina* rootstock _____ 97

A rubber plant that has grown too tall for the house can be air layered to produce a smaller plant. The best time to do this is during late spring or early summer. First make an upward cut into the stem. This cut will exude a white latex.

Hold the cut open by inserting a matchstick or a small twig. Do not worry about the white latex, this is quite normal. But wash your hands thoroughly when the operation is over.

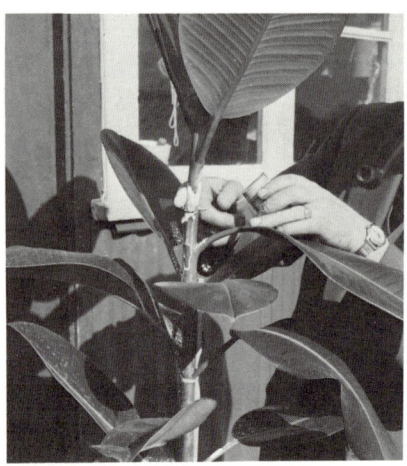

Apply hormone rooting powder to the whole of the cut area.

Moisten a large handful of moss and wrap it round the wounded stem of the rubber plant.

Holding the moss in position wrap a length of polythene around the stem and fasten this securely at the top and bottom to form an airtight sleeve round the wound.

Keep an eye on the layered plant and when a mass of white roots is visible through the polythene it is time to remove the top of the plant. Do this by cutting just below the polythene sleeve.

Remove the polythene to expose the flourishing new root system. The layer can be carefully planted in a pot containing a compost such as John Innes potting compost no. 1 or a peat-based equivalent. The bottom half of the plant will in time produce side shoots which in turn can be air layered.

The tips of blackberry canes will produce roots if they come into contact with the soil in autumn. This characteristic can be used to propagate favourite varieties. Fill a 3½-in pot with a sandy compost and insert the tip of a cane with the aid of a dibber. Firm in with the fingers and secure with a piece of bent wire.

During the following spring remove the layer carefully from its pot. By this time a root system will have developed from which shoots will form. When this has happened the original cane can be severed.

This shows the development of a tip layer. The terminal bud looks quite normal when it first goes into the ground. Soon roots develop from around this bud. Eventually the shoot bud begins to grow (indicated by pencil point) to give rise to a new plant.

Strawberries propagate themselves by long stems, known as runners. First make sure the parent plant is completely healthy, showing no sign of mottled or twisted foliage which could suggest virus disease. Start by sinking 3-in pots into the soil and filling them with potting compost to which some sharp sand has been added.

Next, peg down each runner into a pot with a piece of bent wire. Each runner should only be allowed to produce one plantlet and any extension growth should be cut off. It is advisable to remove the flowers from a plant intended for propagation purposes so all the energy goes into forming the new plantlets.

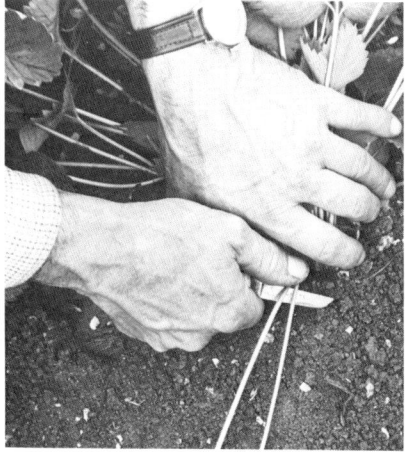

Water each plantlet in using a can fitted with a fine rose. Allow the plants to establish themselves and form roots whilst they are still receiving nutrients from the parent plant through the runner.

At the end of the season the runner can be severed at either end close to the parent and daughter plants. The new plants can then be lifted either for forcing or for establishing a new bed.

Chlorophytum produces a large number of baby plants on runners. These can continue to grow on the parent plant or can be layered to form new plants.

Fill a 3-in pot with sandy compost and peg a baby plant down into the compost with a hair pin. You can often see the beginnings of the roots at the base of the plantlet.

The baby chlorophytum should be kept watered and will soon root. The runner can then be cut, separating the plantlet from the parent plant.

Large bushy plants of heathers (erica) can be propagated by a form of layering known as stooling. This involves taking out a shallow trough of soil around the plant and bending the stems outwards into this depression. The centre of the plant is then covered with sandy compost which is well firmed down. The outermost tips to the branches should not be covered so they can grow upwards.

In time the lower parts of the stems covered with compost will produce roots. All these rooted branches can be separated from the parent plant and lifted.

These new young plants can be replanted in their final growing positions straight away after lifting in autumn or spring.

Budding and Grafting

Grafting involves joining a piece of one plant onto another, entirely separate, growing plant. This latter, known as the stock, provides the root system and the piece grafted on, which is called the scion, will eventually provide the top part of the plant. The stock and scion must be closely related and usually belong to the same genus and, provided they are compatible, will knit together to form a flourishing plant which will have characteristics of both the stock and scion plants. This is the object of grafting: to combine together the desirable qualities of two plants.

Most productive fruit trees are grafted onto a stock. The stock controls the size and vigour of the tree and the scion the type of fruit it will bear. Two examples will make this point clearer. The apple Cox's Orange Pippin grafted onto a Malling 9 rootstock will produce a dwarf tree, usually trained as a cordon, that will need to be staked throughout its life. This is because Malling 9 has a brittle, shallow root system only capable of supporting a small bush. It will bring the tree into bearing early but will not be long lived. On the other hand if the same variety is grafted onto MM111 rootstock the tree will grow tall and vigorously. It will take several years longer before it produces a crop but will be long lived. This is because MM111 if allowed to grow on would itself develop into a large, vigorous tree. It is also resistant to woolly aphid and this characteristic will be carried over to the Cox scion as well.

Grafting has a long and interesting history and there are many methods, but this book will only deal with those most useful to the person with a small garden. Whip and tongue is perhaps the most straightforward method of grafting. The stock and scion must be of the same thickness. It is a method which is best suited to small subjects. The rootstock is produced by stooling as described on page 91 and must be at least a year old. The scion should have been selected and cut in January or February and heeled in beside a north-facing wall or hedge until March or April when grafting can be carried out. Saddle grafting also requires the stock and scion to be of the same size so they fit neatly together. This method is used almost exclusively for grafting varieties of rhododendron onto _Rhododendron ponticum_ rootstocks. In rind or crown grafting the stock is usually mature and considerably thicker than the scions. Several scions can be grafted round the edge of the stock in this method. It is used for reworking trees, most commonly apple, that have become unfruitful or badly neglected.

Budding is another method of grafting fruit trees but is most widely used for propagating varieties of roses. The scion is merely a sliver of bark with one bud. The leaf stalk is usually left in place to facilitate handling such a tiny item. This bud is inserted under the bark of the stock, below soil level for roses, above soil level for a fruit tree or about $4\frac{1}{2}$ ft (1.35 m) above soil level for a standard specimen. As in other methods of grafting, budding is carried out so a particular rose, for instance, with a lovely flower but with weak roots or a straggly habit can be made to grow in a more desirable way. For example, it can be in the form of a bush or as a climber, it may gain a firmer anchorage from an improved root system or have resistance to a certain disease it may normally be very susceptible to.

As with all the other methods of propagation described in this book, hygiene is again of paramount importance. Both stock and scion must be strong and healthy with no signs of any disorder. The knife used, preferably a special budding knife, must be very sharp

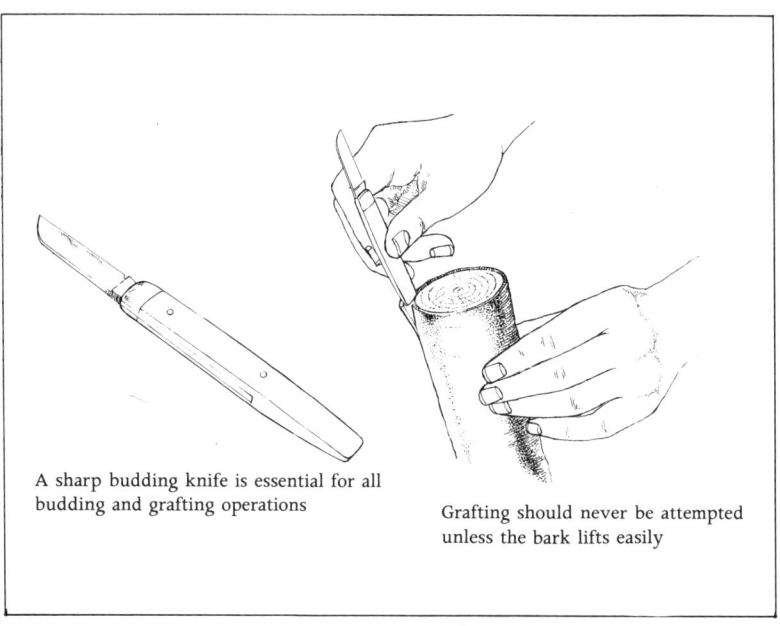

A sharp budding knife is essential for all budding and grafting operations

Grafting should never be attempted unless the bark lifts easily and the blade clean. Grafting operations should only be carried out when the bark lifts easily away from the hardwood after a cut has been made. If this fails to happen then the time is not right and the graft will not take.

Grafting involves the knitting together of the stock and scion and, unless there are incompatibility problems, this can only take place if the cambium layers (those cells which are capable of generating growth and which are situated directly underneath the bark) are held firmly in contact with each other. Soft natural raffia is the best tying material to use. This should be liberally covered with a special grafting wax which will seal the wounded area and prevent rain, pests and diseases from entering. As soon as the scion shows signs of growth the wax and raffia ties should be removed otherwise they may cut into the swollen bark and damage the stock (often called girdling). This is the time to cut back the top growth of the stock in budded specimens, usually to about 4 in (10 cm). The scion can be tied to this 'snag' for support initially.

A successful graft should not be neglected. The ground around the base should be kept clear of weeds and great care should be taken not to damage the stem with the hoe or fork. Any growth which shoots from the rootstock – sucker growth in roses – should be removed, otherwise the stock, often more vigorous than the scion, will take over eventually. These shoots and suckers can usually be recognized quite easily and can always be traced back to the stock. A common example where this has happened and the shoots have not been removed may be seen in flowering cherries which bear both pink and white blossom and single and double flowers. The single white flowers are those produced by shoots from the stock, *Prunus avium*, the wild cherry.

Bush roses are budded on to rootstocks, usually *Rosa canina*, which will form the root portion. When the bud shows signs of growth the top part of the rootstock will be cut off.

Prepare a trench 9 in (23 cm) deep so that the roots can be laid in. Plant stocks in winter at 1-ft (30-cm) spacing to give sufficient working space when budding.

After planting, firm the soil evenly around the rootstocks. If more than one row is being planted, space the rows at 2 ft (60 cm) apart.

The base of the rootstock when the soil has been removed, showing the portion of the stem where the bud is inserted in summer. The original soil level is indicated by the line.

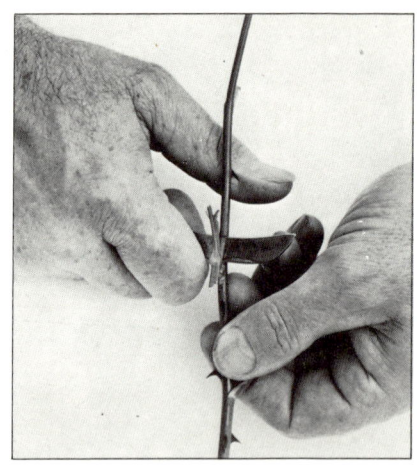

Take buds from young shoots cut from flowering bushes in June or July. The shoots should be stood in a jar of water until required to keep them turgid.

Remove the leaves but not the leaf stalks from each shoot. Detach the buds by slicing beneath them with a sharp knife. Use only stout, well-formed buds.

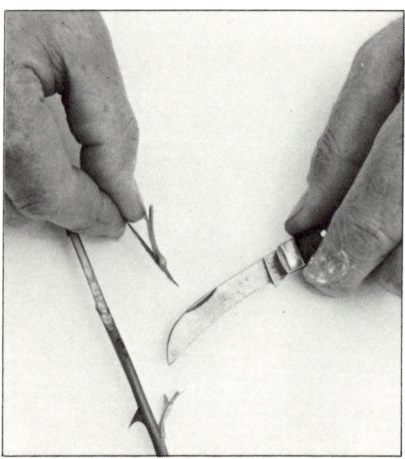

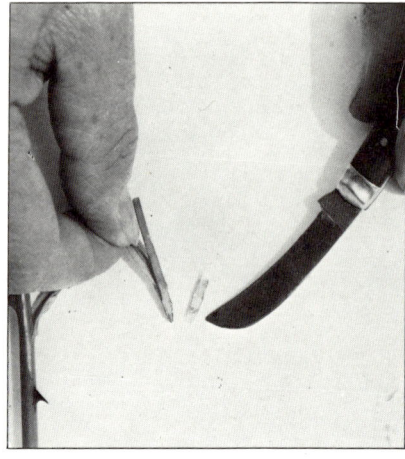

This shows a bud detached correctly. A young shoot may have six or more buds, all of which can be used.

Remove the sliver of wood at the back of the bud cleanly with the tip of the knife. It should come away easily. The intact leaf stalk forms a convenient handle with which to hold the bud.

Make a T cut in the stock at the lowest point possible, after removing 1 to 2 in (2.5 to 5 cm) of soil from around the base of the stem. If the bud is inserted higher up, suckers may result later on.

Form the top of the T first, using a sharp budding knife. Then make a downward cut about 1¼ in (3.5 cm) in length.

Open up the flaps of bark of the T with the handle of the budding knife to make the insertion of the bud easier.

Insert the bud under the flaps of bark of the T cut. Leave the leaf stalk in position as it will drop away after the bud has taken. *(Continued overleaf)*

A very convenient and quick
method of securing the bud is
to use thin rubber patches
with two thin wire spikes.

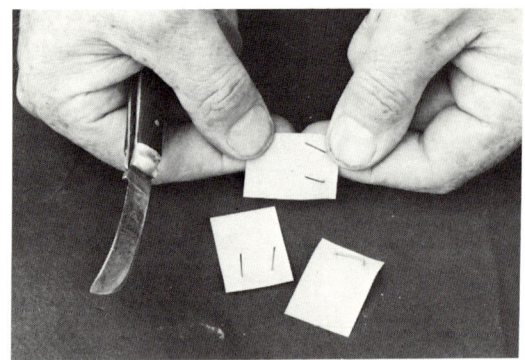

Place the patch in position
over the bud, draw it round
the stem and secure the wire
spikes.

An alternative method is to use moistened
wide raffia. Bind this around the graft
taking care to go above and below the
actual bud.

The bud is still visible when it has been
bound securely in position. When the
bud starts into growth in the spring, slit
the raffia immediately. Cut off the portion
of the rootstock above the inserted bud in
spring so all the energy of the plant goes
into the new growth arising from the bud.

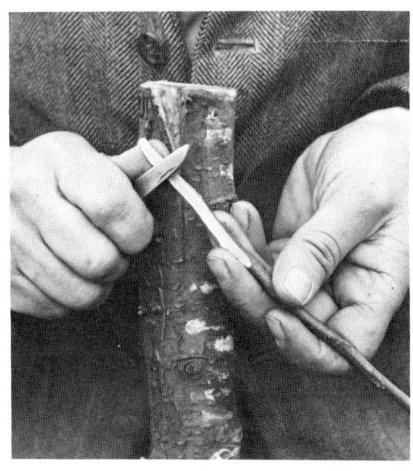

Rind grafting is most often used on mature fruit trees which have grown out of hand through neglect or have become unfruitful. By grafting on young scions the tree can be given a new life. Cut back each branch in winter and collect suitable material for scions, heeling them in in a cool protected place and graft in March or April. First cut a slit in the bark.

The material for the scions is gathered from the desired varieties in winter. Prepare each scion with a sloping cut $2\frac{1}{2}$ to 3 in (6 to 8 cm) long.

Insert each scion in a slit in the bark: two opposite each other in a branch 2 or 3 in (5 or 8 cm) in diameter, more in a thicker branch.

After insertion tie the slit area of the branch round firmly with raffia to keep the scions secure. *(Continued overleaf)*

Cover the tied raffia with grafting wax, which may be cold or of a type which requires heating.

The finished appearance of a bush tree on which three branches have been re-worked or re-headed by the rind method.

The scions unite with the branch at the point of union of the two cambium layers, and buds send out new shoots. Note that the raffia has been severed; this would have been done as soon as growth on the scions was apparent – usually about a month after grafting had taken place. If this is not done damage could be caused to the tree by girdling.

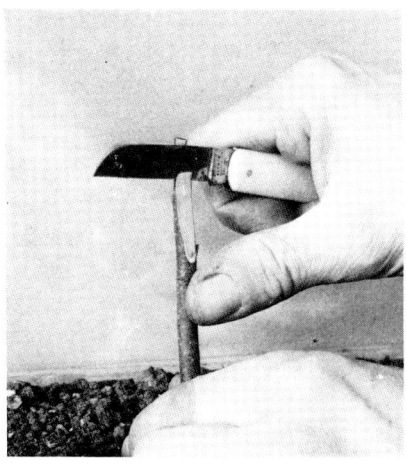

This method is used where the stock and scion are of approximately the same size. The rootstock must be at least one year old. Make a long slanting cut through the rootstock, removing the top of the plant. Then make a nick in the cut face of the slanting cut to form the tongue.

The scion should be the previous year's growth cut in January or early February and heeled in in a protected place until late March or April when grafting takes place. The base of the scion is cut in a similar way to correspond with the stock.

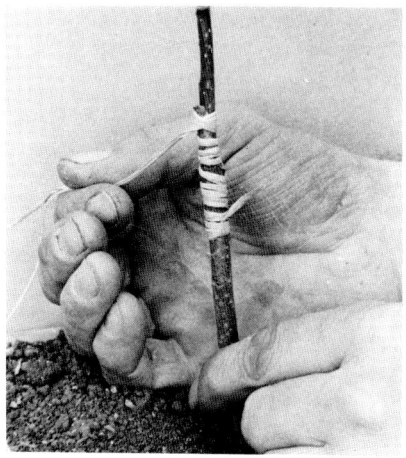

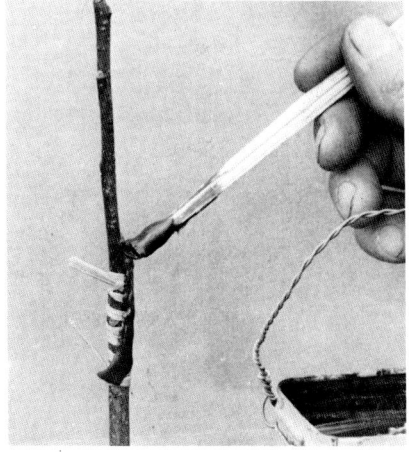

Place the scion in position on the stock so the tongues interlock holding the two parts to be grafted together. They are then bound with soft raffia so the cambium layers of each are in contact.

The whole cut area of the graft is now generously painted with a good grafting wax to keep out all moisture, pests and diseases. When the scion starts to shoot, the wax and raffia can be carefully cut through to prevent girdling.

This method of grafting is used almost exclusively for rhododendrons. The scion is prepared by making two upward slanting cuts in the base of the stem as shown.

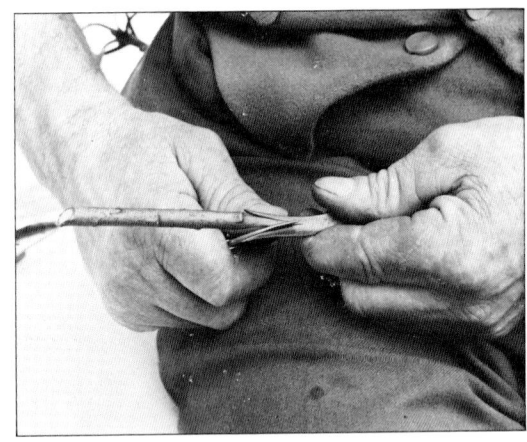

The stock, usually *Rhododendron ponticum*, is prepared by making two upward slanting cuts that meet in a point. The stock then sits neatly on this 'saddle' and is bound in position with soft raffia so the cambium layers of stock and scion are held in contact. Grafting wax is then painted over the whole area. When the scion shoots the wax and raffia should be cut to prevent girdling.

Grafting cacti is worth trying when plants are available. With a sharp knife remove the top cleanly from the chosen stock plant, which must be strong growing, and remove a piece from a slow-growing variety so the two cut surfaces match as nearly as possible.

Place the new top piece carefully in position on the stock plant. As it is imperative that it does not move, place a piece of string with a lead weight attached to each end over the top and allow to hang as shown.

To form a vigorous Christmas cactus insert a leaf or scion into a slit made in a suitable stock, such as cereus. Pin in place with a cactus thorn.

When the scion is in position, remove the top from the stock plant. This ensures that all nutriment goes to develop the new scion.

Bulbs and Corms

Bulbs and corms have been allocated a separate chapter as their mode of growth is quite different from that of the other plants dealt with in this book. Both bulbs and corms can be grown from seed but in many cases this is a long process and it can be six or seven years before a daffodil raised in this way will flower. So it is very often preferable to propagate them by vegetative means. The methods of propagation mostly fall into the category of division, for bulbs and corms can both be divided by cutting them into pieces and by removing offsets. However, there are differences that must be borne in mind when dealing with these types of plants.

Bulbs and corms are composed of swollen underground leaves or stems that act as food stores and sustain the plants during unfavourable periods. These underground storage organs enable the plants to send up leafy growth and flowers as soon as conditions above ground begin to improve. The leaves then manufacture food which is passed down to develop a new storage organ and so the cycle continues from year to year. As well as developing a main

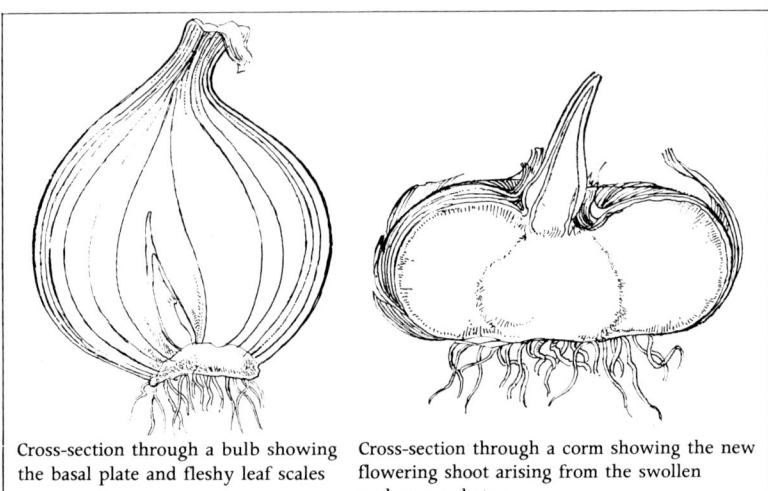

Cross-section through a bulb showing the basal plate and fleshy leaf scales

Cross-section through a corm showing the new flowering shoot arising from the swollen underground stem

replacement food store small bulbs and corms may be formed. There may be only one or two, known as offsets, or there may be many, variously referred to as spawn, cormels or bulblets according to the plant.

It is easier to deal with bulbs and corms separately as there are a number of basic differences between them. The main storage organ in bulbs is formed from fleshy scale-like leaves (frequently just the thickened leaf bases) and which may either be covered with a papery membrane, as in narcissus, tulip and onion, or be quite naked, as in lilies and fritillarias. Bulbs are often planted and then left alone, but most, if they are lifted every few years after the foliage has died down, will be seen to have produced offsets of various sizes. The offsets can be separated from the bulb clusters and should be grown on for a year or two in a nursery bed to build up their store of energy before they are planted into their flowering quarters.

Both types of bulbs produce offsets but the naked bulbs can also be reproduced from scales. This involves breaking off the scales and mixing them with damp compost. They will show signs of root formation in a few weeks if kept in a warm place and this is the signal for planting them out in pots or trays. Stem-rooting lilies can be encouraged to produce small bulbs in the axils of their leaves if their flowers are removed just before they open. These bulbils can be removed and planted in a seedtray to grow on. In both these cases the small bulbs must be grown on for at least a year to increase in size before they can be planted in the open.

Narcissus, hyacinth and muscari are examples of bulbs with papery skins that can be cut across the basal plate in order to encourage the production of tiny bulbs. Shallow, V-shaped cuts are made across the base and the wounded bulb is then kept in a warm, dark place until the bulblets appear along the edges of the cuts. These bulblets are detached and then grown on for two or three years before they are mature enough to be planted outside in their flowering positions.

Corms concentrate their food store in a swollen underground stem. They too produce offsets but in a different way to bulbs. The old shrivelled corm responsible for the past year's growth can usually still be seen at the base of the new main corm. It should be pulled off and discarded. Between the new and old corm there will be a mass of offsets. Known as spawn or cormels, they have developed during the growing period and are a typical phenomenon of gladiolus varieties. These can be detached and stored in a cool, dry but airy place. In spring they can be planted in a special bed outside and grown on for two years when they will have reached flowering size. Corms can also be cut into three or four pieces, each of which must possess a bud, and after dusting with a fungicide they can be planted out to grow on for at least a year.

Bulbs and corms that are too tender to survive the winter or need to be divided can be lifted after the foliage has turned yellow or died down. Then they must be dried off, preferably on racks of chicken wire in the sun, otherwise in a warm, dry, airy place before the offsets are separated. The bulbs should all be checked over to see they are undamaged and free from disease and then dusted with a fungicide if they are to be stored for the winter. They can then be labelled, packed in boxes and stored in a cool, dry place until planting time.

BULB PLANTING CHART

	Planting Time	Planting Depth	Distance Apart	
Chionodoxa	Sept-Oct	3 in (8 cm)	6 in (15 cm)	bulb
Colchicum	July-Aug	2 in (5 cm)	6 in (15 cm)	corm
Crocus *(autumn flowering)*	July-Aug	3 in (8 cm)	4-6 in (10-15 cm)	corm
(spring flowering)	Sept-Oct	3 in (8 cm)	4-6 in (10-15 cm)	corm
Fritillaria *(Crown Imperial)*	Sept-Oct	4-6 in (10-15 cm)	18 in (45 cm)	bulb
Galanthus	Aug-Sept	4 in (10 cm)	3-6 in (8-15 cm)	bulb
Gladiolus	Mar-May	3 in (8 cm)	6 in (15 cm)	corm
Hyacinth	Oct-Nov	3-4 in (8-10 cm)	8 in (20 cm)	bulb
Iris *(bulbous)*	Sept	3-4 in (8-10 cm)	6 in (15 cm)	bulb
Leucojum *(spring flowering)*	Sept	3 in (8 cm)	3-6 in (8-15 cm)	bulb
(autumn flowering)	July-Aug	3 in (8 cm)	3-6 in (8-15 cm)	bulb
Lilium *(stem rooting)*	Dec-Mar	4-6 in (10-15 cm)	6-8 in (15-20 cm)	bulb
(not stem rooting)	July-Aug	1-3 in (2-8 cm)	6-8 in (15-20 cm)	bulb
Muscari	Sept-Oct	2-3 in (5-8 cm)	4-5 in (10-13 cm)	bulb
Narcissus	Sept-Oct	5 in (13 cm)	4-5 in (10-13 cm)	bulb
Scilla	Aug-Oct	2-3 in (5-8 cm)	3-6 in (8-15 cm)	bulb
Tigridia	Apr-May	3 in (8 cm)	4-6 in (10-15 cm)	corm
Tulipa	Oct-Nov	4 in (10 cm)	4-6 in (10-15 cm)	bulb

Every four or five years naturalised daffodil bulbs can be lifted, usually round about June, after the foliage has yellowed and begun to die down. It will be seen that each bulb has produced one or more offsets. These can be gently separated.

The bulbs and offsets should be dried in the sun and stored in a dry place until August. They can then be planted outside in their flowering positions again.

Ideally tulips should be lifted annually when the foliage has died down. Then any offsets that have developed can be separated off. These will probably be quite small and unless you want to grow them on for two years they can be discarded. If tulips are left in the ground for a longer period the offsets will be more numerous and larger. These larger ones can be separated and replanted in the autumn.

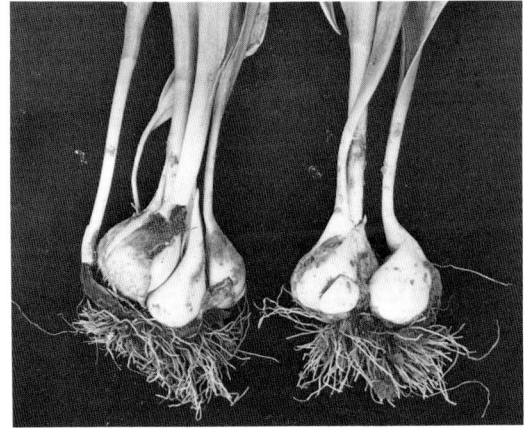

Gladiolus corms produce a mass of small offsets in most years. These are known as spawn or cormels. After lifting in the autumn the spawn will be clearly visible at the base of the new main corm.

The spawn can be removed and dried and stored over winter just like the main corms. The latter should be inspected for any sign of damage or disease, dusted with fungicide, labelled and stored in boxes in a cool dry place. The shrivelled remains of the old corm should also be removed and can be discarded.

In spring the cormels can be planted outside to grow on, then lifted in the autumn again. The following year this growing on should be repeated as they usually take two seasons to reach flowering size.

Bulbs, such as hyacinths, produce offsets very slowly in the wild unlike narcissus and tulips which usually divide during a year. So one method of encouraging the formation of offsets is to score across the basal plate. Take out two small wedges of tissue at right angles to each other. Dust these cut surfaces with a fungicide such as flowers of sulphur or captan and rest the bulb, cut side up, on a bed of sand in a warm dry place. An airing cupboard would be ideal.

After about two months some small bulblets will have grown from the cut surfaces. These will vary both in size and number.

The bulblets can be carefully removed with the fingers and planted in compost in a seed-tray to be grown on to flowering size. In the case of the larger bulblets this will be one year, the smaller ones will take two years before they can be planted outside.

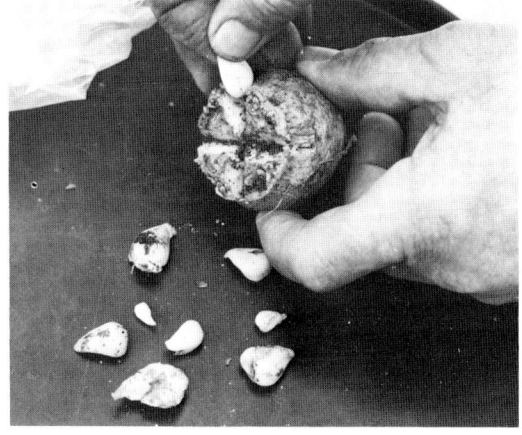

Lily bulbs are not covered by the papery layer that protects narcissus and tulips so their scales are clearly visible.

These scales can be carefully removed from a bulb for propagation purposes, but take care different varieties do not get mixed up.

The scales are mixed in a damp peat into which some grit or sharp sand and a little fungicide, such as flowers of sulphur or captan, has been incorporated.

Put the peat mixture and the lily-bulb scales into a polythene bag and fasten this at the top with a plastic-covered tie. Place the bag into the airing cupboard or another warm, dark place.

Inspect the scales regularly and when they show signs of shooting remove them from the airing cupboard and tip them out of the bag. The scales should then be planted out in a seedtray to develop one or more tiny bulblets at their bases.

Carefully remove the bulblets from the scale, which can then be discarded. The bulblets should be planted out in a seedtray to be grown on for a season before being hardened off and planted outside.

Stem-rooting lilies, such as *Lilium tigrinum* and *L. speciosum*, produce tiny bulbs known as bulbils in the axils of the leaves. Production of these bulbils is encouraged if the flower buds are removed just before they open.

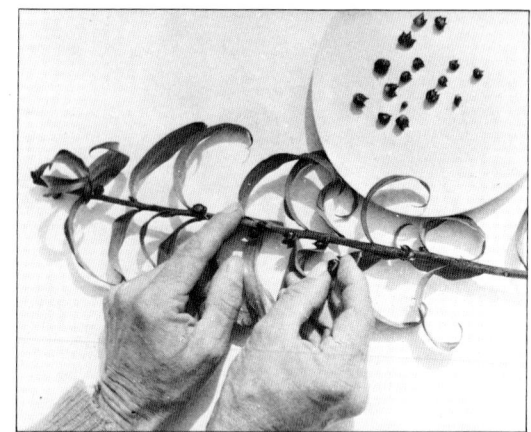

The bulbils are removed from the leaf axils as they develop and are sown in a seedtray in a peat-based compost. Space these tiny bulbs about 2 in (5 cm) apart in each direction and approximately 1 in (2.5 cm) deep.

The young plants produced from the bulbils are potted up individually into 3½-in pots filled with a peat-based compost. Grow on under glass and plant out in the garden the following autumn. When planting, place each bulb on a layer of sharp sand to assist drainage.

Plant List

The following alphabetical list covers a number of the plants that can be propagated with success by the home gardener. It is by no means comprehensive but it does give an idea of the wide range of plants that can be raised successfully with just the basic equipment and know how. The page numbers referred to give details of propagating that specific plant or, if it has not been mentioned by name, to the methods to be followed and to any other information that could be helpful.

Acer (maple) Seed (pages 10-31), should be stratified (page 22); bud variegated and Japanese varieties in summer (pages 104-6 and 108-110); layer in autumn (page 93).

Achillea (yarrow) Seed (pages 10-31); division (pages 80-2).

Allium (ornamental onions) Seed (pages 10-31); divide offsets (page 119).

Annuals – half-hardy Seed (pages 10-31).

Annuals – hardy Seed (pages 10-31).

Apple Bud or graft (pages 104-113).

Bamboo Seed (pages 10-31); division (pages 82 and 84).

Begonia – fibrous rooted Seed (pages 10-31); division (pages 80-2).

Begonia – foliage varieties Seed (pages 10-31); leaf cuttings (page 67).

Begonia – tuberous rooted Seed (pages 10-31); division (page 87).

Berberis (barberry) Seed (pages 10-31); layering (page 93).

Blackberry (rubus) Tip layering (page 100).

Buddleia Seed (pages 10-31); cutting: (pages 41, 58-9 and 62).

Cactus Seed (page 23); grafting (page 115).

Camellia Seed (page 10-31); leaf-bud cuttings (page 66).

Chaenomeles Seed (pages 10-31), should be stratified (page 22); division of suckers (page 89); layering (page 93).

Chlorophytum (spiderwort) Division (pages 80-2); layering (page 102).

Chrysanthemum Seed (pages 10-31); cuttings (pages 42-3).

Clematis Seed (pages 10-31); layering (page 94).

Coleus Seed (pages 10-31); cuttings (pages 39-41)

Conifers Seed (pages 10-31); cuttings (page 61).

Convallaria (lily-of-the-valley) Division (pages 80-1).

Cyclamen Seed (pages 10-31).

Dahlia Cuttings (pages 44-5); division of tubers (page 87).

Delphinium Seed (pages 10-31); cuttings (page 46); division (pages 80-2).

Dianthus (pinks and carnations) Seed (pages 10-31); cuttings (page 55); layering (page 92).

Dracaena Seed (pages 10-31); root cuttings (page 73); air-layering (pages 97-8).

Escallonia Seed (pages 10-31); cuttings (pages 41, 58-9 and 62).

Ferns Division (pages 80-2); spores should be sown in sandy compost and left uncovered but shaded in a cold frame.

Forsythia Cuttings (pages 41, 58-9 and 62); layering (page 93).

Fuchsia Seed (pages 10-31); cuttings (page 47).

Gaillardia Seed (pages 10-31); root cuttings (pages 74-6).

Galanthus (snowdrop) Division of offsets (page 119); scoring (page 121).

Gladiolus Division of offsets (page 120).

Gooseberry Cuttings (page 65).

Heathers (calluna, erica) Cuttings (page 61); layer by stooling (page 103).

Hebe Seed (pages 10-31); cuttings (pages 58-9).

Hedera (ivy) Cuttings (pages 41, 58-9 and 62).

Hosta Division (pages 80-2).

Hydrangea Cuttings (page 48).

Impatiens (buzy lizzie) Seed (pages 10-31); cuttings (pages 41).

Iris – bulbous Division of offsets (page 119).

Iris – rhizomatous Division (page 85).

Jasminum Cuttings (pages 41, 58-9 and 62); layering (page 93).

Laurus nobilis (bay laurel) Cuttings (page 62); layering (page 93).

Lavandula (lavender) Cuttings (page 60).

Ligustrum (privet) Seed (pages 10-31); cuttings (page 41).

Lilium (lilies) Seed (pages 10-31); division of offsets (page 119); from scales (pages 122-3); stem bulbils from stem-rooting lilies only (page 124).

Lonicera (honeysuckle) Seed (pages 10-31), should be stratified (page 22); cuttings (pages 58-9 and 62); layering (page 93).

Lupinus (lupins) Seed (pages 10-31); cuttings (page 49); division (page 82).

Magnolia Seed (pages 10-31), should be stratified (page 22); cuttings (page 57); layering (page 93).

Meconopsis (Himalayan and Welsh poppies) Seed, sow as soon as ripe (pages 10-31).

Myosotis (forget-me-not) Seed (pages 10-31).

Narcissus (including daffodils) Division of offsets (page 119); scoring (page 121).

Nerine (Jersey lily) Seed (pages 10-31); division of offsets (page 119).

Mint Culinary mint can be forced in winter in a warm place from root cuttings (page 79).

Paeonia (paeony) Seed (pages 10-31); division of herbaceous species (page 84); tree species should be layered (page 93).

Papaver (poppy) Seed (pages 10-31); perennial species from root cuttings (pages 74-6).

Pear Bud or graft (pages 104-113).

Pelargonium (bedding geraniums) Seed (pages 10-31); cuttings (pages 50 and 53-4).

Peperomia Cuttings (page 68).

Phlox Seed (pages 10-31); cuttings (page 79); division (pages 80-2).

Prunus (ornamental cherries, peaches and almonds) Seed (pages 10-31), should be stratified (page 22); grafting or budding (pages 104-113).

Rhododendron Seeds (pages 10-31); layering (page 93); grafting (page 114).

Rhus (sumach) Layering (page 93); division of suckers (page 89).

Ribes (fruiting and flowering currants) Cuttings (page 64); stooling (page 103).

Romneya Seed (pages 10-31); root cuttings (pages 74-6).

Rosa (roses) Species can be raised from seed (pages 10-31) which should be stratified (page 22); cuttings (page 57 and 62); budding (pages 107-110).

Rudbeckia Seed (pages 10-31); division (pages 80-2).

Saintpaulia (African violet) Seed (pages 10-31); leaf cuttings (page 69).

Sansevieria (mother-in-law's tongue) Division (page 88).

Saxifraga Seed (pages 10-31); division (pages 80-1 and 86).

Sedum Seed (pages 10-31); leaf cuttings (page 71); division (pages 80-1 and 86).

Sempervivum Seed (pages 10-31); division (pages 80-1 and 86).

Sorbus (mountain ash) Seed (pages 10-31, should be stratified); graft or bud varieties (pages 104-113).

Strawberry Alpine strawberries from seed (pages 10-31); division (pages 80-1 and 86); layering (page 101).

Streptocarpus Seed (pages 10-31); leaf cuttings (page 70).

Syringa (lilac) Seed (pages 10-31); cuttings (pages 41 and 57); division of suckers (page 89); layering (page 93).

Thymus (thyme) Seed (pages 10-31); cuttings (page 60); division (pages 80-2).

Trollius (globe flower) Seed (pages 10-31); division (pages 80-2).

Tulipa (tulip) Division of offsets (page 119).

Viburnum Seed (pages 10-31), should be stratified (page 22); cuttings (page 57); division of suckers (page 89).

Viola (including pansies) Seed (pages 10-31); cuttings (pages 38-40 and 47); division (pages 80-1 and 86).

Vitis (vine) Seed (pages 10-31); eye cuttings (page 72).

Weigela Cuttings (pages 58-9); layering (page 93).

Wisteria Seed (pages 10-31); eye cuttings (page 72), layering (page 93).

Yucca Root cuttings (pages 74-6); division of offsets (pages 80-1 and 86).

Zantedeschia (arum lily) Division (pages 80-1); remove small rhizomes in late summer and these will grow into new plants.

Index